1st METECH Workshop – From Deep-Sea to Coastal Zones: Methods and Techniques for Studying Paleoenvironments 2009

IOP Conference Series: Earth and Environmental Science
Volume 5

Faro, Portugal
25 - 29 February 2009

ISBN: 978-1-5108-0205-6
ISSN: 1755-1307

1st METECH Workshop – From Deep-Sea to Coastal Zones: Methods and Techniques for Studying Paleoenvironments 2009

IOP Conference Series: Earth and Environmental Science
Volume 5

Faro, Portugal
25 - 29 February 2009

ISBN: 978-1-5108-0205-6
ISSN: 1755-1307

IOP Institute of Physics

Printed from e-media with permission by:

Curran Associates, Inc.
57 Morehouse Lane
Red Hook, NY 12571

Some format issues inherent in the e-media version may also appear in this print version.

Printed by Curran Associates, Inc. (2015)

For permission requests, please contact the Institute of Physics
at the address below.

Institute of Physics
Dirac House, Temple Back
Bristol BS1 6BE UK

Phone: 44 1 17 929 7481
Fax: 44 1 17 920 0979

techtracking@iop.org

Additional copies of this publication are available from:

Curran Associates, Inc.
57 Morehouse Lane
Red Hook, NY 12571 USA
Phone: 845-758-0400
Fax: 845-758-2634
Email: curran@proceedings.com
Web: www.proceedings.com

Table of contents

Volume 5

FROM DEEP-SEA TO COASTAL ZONES: METHODS AND TECHNIQUES FOR STUDYING PALEOENVIRONMENTS
25–29 February 2008, Faro, Portugal

Accepted papers received: 6 January 2009
Published online: 10 February 2009

MICROPALEONTOLOGICAL PROXIES

ISOTOPIC PROXIES

GEOCHEMICAL PROXIES

SEDIMENTOLOGICAL PROXIES

Dynamics of living phytoplankton: implications for paleoenvironmental reconstructions

Ana B Barbosa

Centre for Marine and Environmental Research (CIMA), Universidade do Algarve, Campus de Gambelas, 8005-139 Faro, Portugal

E-mail: abarbosa@ualg.pt

Abstract. Phytoplankton is the dominant primary producer in aquatic ecosystems and is considered a gauge of ecological condition and change. Some phytoplankton groups, namely diatoms, dinoflagellates, and coccolithophores, produce morphological or chemical fossils that can be used for paleoenvironmental reconstruction. This study aims to review the processes that regulate dynamics in living phytoplankton and to highlight how this knowledge is used in paleoecological studies. The distribution patterns of phytoplankton in present-day aquatic ecosystems are shaped by the interplay between processes that regulate cell growth and cell death. Cell growth and cell death are regulated by the internal environment of phytoplankton (e.g., specific environmental tolerances, resource uptake properties, cell size, density and morphology, alternative nutritional strategies such as mixotrophy or N_2 uptake, motility, intracellular storage capacities, grazing resistance properties), and by its external environment. The external environment includes variables dependent on the availability of resources (e.g., light intensity, concentration of CO_2 and dissolved inorganic macronutrients and micronutrients, availability of living prey in case of mixotrophs) and variables independent of resources (e.g., temperature, salinity, turbulence, ultraviolet radiation, bioactive compounds, activity of grazers, viruses, and eukaryotic parasites). The importance of recently described loss processes, such as grazing by phagotrophic protists, viral lyses, and programmed cell death, is discussed in the context of its potential impact upon phytoplankton vertical fluxes. Examples of the use of different phytoplankton metrics (e.g. abundance, species composition, species morphology, and elemental composition) to infer contemporaneous as well as past environmental and ecological conditions are critically evaluated.

1. Introduction

Phytoplankton constitutes a heterogeneous group of unicellular photosynthetic organisms, both prokaryotic and eukaryotic, that drift freely in the water column. Phytoplankton is the dominant primary producer in most aquatic ecosystems and represents almost 50% of net primary production on Earth. Besides its critical function of primary production, that directly and indirectly fuels aquatic food webs, it has significant impact on water quality and is a key player in global biogeochemical processes, mediating cycling, sequestration, and exportation of inorganic and organic compounds. Since phytoplankton is highly sensitive to environmental changes it is also extensively used as a gauge of ecological condition and change. Overall, phytoplankton is of paramount importance to understand and predict the carrying capacity of aquatic systems, the production of living aquatic resources, and the responses to natural and anthropogenically induced environmental changes [1, 2, 3, 4, 5].

Although most phytoplankton cells lack substantial hard parts and rarely fossilize, some can leave preservable and recognizable physical remains constituting organic-walled or mineral-walled microfossils (coccolithophores, calcitic-walled dinoflagellates cysts, diatoms, silicoflagellates, and chrysophytes). Thus, the fossil record of phytoplankton is inevitably incomplete and obviously biased toward groups that produce mineralized skeletons or decay-resistant organic walls [6, 7]. However, some phytoplankton cells can also produce diagnostic biomarker molecules (or chemical fossils), that are usually associated to photosynthetic pigments or phospholipid-derived fatty acids from cell membranes. These biomarker molecules can be related either directly or indirectly, through a series of diagenetic alterations, to their biogenic sources. Chemical fossils are particularly valuable in case of organisms that do not leave recognizable morphological fossils and can be used as proxies for total and group-specific phytoplantkon biomass or paleoproductivity [7, 8, 9, 10, 11]. Phytoplankton derived fossils, typically combined with other geological evidences and complementary proxies, are useful tools for reconstruction of paleoaquatic environments. They have been used to infer past physical and chemical environmental conditions [6, 9, 10, 12, 13, 14], as well as attributes of past biological communities [10, 11, 12, 13, 14, 15, 16]. Thus, phytoplankton derived fossils allow the study of both natural and anthropogenic processes, occurring at local and global scales (e.g., anthropogenic eutrophication, industrial pollution, coastal upwelling, climate change), their temporal variability at short- to long-time scales, and their impact upon biota [14, 15, 16, 17, 18, 19, 20].

The application of phytoplankton derived fossils to paleoenvironmental reconstruction should evidently take into consideration all the processes that control fossil distribution patterns. Distribution of phytoplankton derived fossils in sediments mainly depends on phytoplankton production rates in the water column, and post-mortem events associated to transport and preservation processes, occurring both at predepositional and postdepositional stages [6]. Hence, knowledge on the processes that regulate the production and dynamics of living phytoplankton can improve our capacity to extract information and interpret microfossil distributions. In this context, the aims of this study are to examine the processes regulating dynamics of living phytoplankton in aquatic ecosystems, highlighting taxa with applications to paleoecology, and to explore how this knowledge is used in the assessment of past environmental conditions. This study will briefly overview aspects of morphology, physiology, and behaviour of phytoplankton, emphasizing those with high impact upon phytoplankton accumulation is sediments, it will then survey the variables and processes that regulate both *in situ* growth rates and *in situ* loss rates of phytoplankton, and it will subsequently highlight links between dynamics of living phytoplankton and the exploration and interpretation of fossil records.

2. Phytoplankton: a biological overview with reference to its paleoecological detection

Phytoplankton comprises a diverse, polyphyletic assemblage of single-celled algae, extremely variable on the level of cell ultrastructure, taxonomic affiliation, cell size, behaviour, and physiology, that use sunlight as an energy source for photosynthesis. Like most photoautotrophic organisms, phytoplankton additionally uses dissolved CO_2 as a carbon source, and inorganic dissolved compounds as nutrients. Living phytoplankton is therefore largely restricted to the photic zone of the water column. All phytoplanktonic living cells have chlorophyll *a*, a key pigment usually used as a proxy for phytoplankton biomass, as well as a series of accessory pigments. Some accessory pigments are ubiquitous (e.g., β-carotene), whereas others are shared by some groups (e.g., diadinoxanthin) or exclusive of specific groups (e.g., alloxanthin, peridinin) and thus can be used as taxonomic biomarkers. Phytoplankton includes several ecologically distinct groups that include cyanobacteria, diatoms, dinoflagellates, and other nano- and microplanktonic flagellates [21].

2.1. Cyanobacteria

Cyanobacteria or blue-green algae are prokaryotic organisms with phycobilins as accessory pigments, and include both coccoid solitary forms and colonial and filamentous forms. The importance of picoplanktonic (0.2–2.0 μm) cyanobacteria has been recognized relatively recently, but this group is clearly considered the dominant phytoplankton group in most aquatic ecosystems and the most abundant photosynthetic organisms in Earth. Due to their relatively high surface area to volume ratio, cyanobacteria are usually considered well adapted to low resource availability. Moreover, some species can fix N_2, further alleviating a potential nutrient limitation. Other ecologically relevant features of cyanobacteria include gas vacuoles, that can regulate cell buoyancy allowing vertical displacements in the water column, high resistance to sinking mostly due to reduced cell size, high resistance to predation, and production of potent toxins and specific photoprotective compounds [21, 22, 23]. Filamentous and mat associated benthic taxa, but not planktonic coccoid groups, have an extensive microfossil record [6]. However, the use of specific biomarker compounds (photosynthetic pigments, photoprotective compound scytonemin, lipid biomarkers) or their geologically stable derivatives, combined with paleogenomical tools, can also be used to trace past presence or biomass of cyanobacteria [9, 24, 25].

2.2. Diatoms

Diatoms are eukaryotic algae with a siliceous bi-valved frustule and intricate architecture, and include both solitary and colonial species with cell sizes usually ranging from 5 to 2000 µm. Diatoms comprise two main groups distinguished by frustule morphology and mode of sexual reproduction, Centrales and Pennales. Centric diatoms are radially symmetric, non-motile, and usually planktonic, while pennate diatoms are bilaterally symmetric, usually benthic, and include both non-motile (araphid) and motile (raphid) species. Diatoms have elevated potential growth rates and, contrary to cyanobacteria, are usually well adapted to high availability of resources, including light and nutrients, and intense turbulence. In fact, diatoms are commonly classified as an opportunistic group, and are the major contributors to spring phytoplankton blooms representing approximately 25% of overall primary production. Under adverse environmental conditions, living diatoms usually sink to the sediment but the occurrence of resting vegetative and spore stages in some neritic species allows extended survival [6, 21, 26, 27]. Diatom remains deposited on the sediment, under favourable conditions, may fossilize leaving recognizable morphological fossils, composed by siliceous frustules, valves or fragments [6]. Fossil pigments (e.g., diatoxanthin), specific lipid biomarkers, and paleogenetic analysis of diatom-derived fossil DNA can also be used as proxies of total or group-specific past presence, abundance, and productivity [6, 10, 11, 15, 28].

2.3. Dinoflagellates

Dinoflagellates are eukaryotic organisms with an external pellicle underlain by amphiesmal vesicles and a special form of eukaryotic nucleus (dinokaryon), and planktonic forms mostly include solitary species. Dinoflagellates include strictly heterotrophic taxa without chloroplasts (aplastidic), strictly autotrophic taxa with chloroplasts (plastidic), as well as mixotrophic taxa that combine phototrophy and phagotrophy. Yet, only plastidic taxa should be considered as components of phytoplankton. Dinoflagellates are divided into armored rigid species, that have vesicles filled with thick cellulose plates, and unarmored flexible species that lack these plates. Most dinoflagellates possess two flagella and are extremely motile, reaching swimming velocities up to 1500 $\mu m.s^{-1}$. This "swim strategy", clearly contrary to the sink strategy exhibited by diatoms, is responsible for daily vertical migrations displayed by many dinoflagellate vegetative cells. Dinoflagellates are particularly sensitive to water column turbulence and are usually adapted to live under low resource availability. This is partly explained by their ability to migrate vertically thus taking advantage of elevated light intensities in surface waters during the day, and increased nutrient concentrations at deeper waters during the night [6, 21, 29, 30]. Moreover, mixotrophic dinoflagellates can thrive under low nutrient, low light conditions through the ingestion of planktonic living prey that sometimes include their own direct competitors [31, 32]. Other ecologically relevant features of dinoflagellates include low potential growth rates [33], elaborate morphological extensions (e.g., horns, spines, crests) that lead to increased nutrient uptake, buoyancy, and predation resistance, and allelopathic, anti-predation defense mechanisms [29, 30]. Following sexual reproduction, at least 10% of all dinoflagellates develop into resting cysts (or dinocysts), composed of highly resistant organic matter, that settle on the sediment.

The induction of sexuality and encystment is apparently associated to a set of different environmental (e.g., nutrient limitation, density discontinuities, other infochemicals) or internal cues, and sometimes cysts are produced even if external nutrients are not limiting [34]. Almost all fossil dinoflagellates are preserved as cysts (organic- and calcitic-walled dinocysts). Thus, although dinocysts represent only a relatively small fraction of original populations of dinoflagellates, these life stages are particularly relevant in paleoecological studies [35]. Moreover, molecular biomarkers of dinoflagellates (e.g., dinosterol) can also be used as proxies for their past presence and biomass [7, 11].

2.4. Coccolithophores and other eukaryotic phytoplankton
Coccolithophores are prymnesiophytes, that usually have two flagella of equal length and a haptonema, with an external shell composed by calcareous scales called coccoliths. These structures may be involved in functions such as protection, buoyancy regulation, and light regulation. Some coccolitophorid species have a life cycle with alternation between a motile and a non-motile planktonic or benthic stage. Coccolitophorids are usually nanoplanktonic (2 – 20 μm) and include both strictly autotrophic and mixotrophic taxa that combine photosynthesis with phagocytosis. Usually, coccolithophores are well adapted to low to moderate nutrient concentrations and high light intensities; yet, some species (e.g., *Florisphaera profunda*) are usually referred as lower-photic-zone-species. Coccolithophore adaptation to low nutrient levels may be explained by phagocytosis of living prey, the use of dissolved organic nitrogen compounds, or to its exceptional phosphorus acquisition capacities. After cell death, coccolithophores sink to the bottom and, under favourable conditions, constitute nanocalcareous fossils composed by individual coccoliths or, occasionally, complete coccosheperes [6, 36, 37, 38]. Coccolithophores also produce specific compounds (e.g., photosynthetic pigments, alkenones) that may constitute chemical fossils or biomarkers and are used as proxies for coccolithophorid presence or biomass [39].

Phytoplankton comprises a series of other organisms including non-motile and flagellated taxa. These algae depict extremely variable cell morphologies, sizes, life cycles, and taxonomic affiliations [46]. Some groups can also produce recognizable morphological microfossils (e.g., internal siliceous skeleton of silicoflagellates, siliceous cysts of chrysophytes, calcareous phycoma of prasinophytes) or specific fossil pigments (e.g., alloxanthin in cryptophytes) [6, 36].

3. Dynamics of living phytoplankton assemblages
The distribution patterns of phytoplankton and its temporal variability in aquatic ecosystems basically reflect the interplay between two modes of regulation: regulation of phytoplankton growth rates (or bottom-up regulation), and regulation of phytoplankton abundance (or top-down regulation). The former mode essentially involves the effects of phytoplankton resources upon cell replication rates, while the later basically includes the processes responsible for phytoplankton mortality or loss in aquatic ecosystems.

3.1. Regulation of phytoplankton growth rate
Phytoplankton instantaneous growth rates or cell replication rates are controlled by several variables but most studies particularly highlight the importance of light intensity, as the energy source, and the concentration of dissolved inorganic macronutrients (N, P, and Si in case of diatoms). Recently, and starting after the “iron revolution” in biological oceanography, the importance of the bioavailability of dissolved inorganic micronutrients (e.g., Fe, Zn, Mg) has been also addressed. Additionally, the impact of CO_2 concentration on phytoplankton growth, particularly relevant to predict its response to global climate changes, has been addressed as well [21, 40]. The relationship between phytoplankton growth rate and the concentration of specific resources is usually described as a hyperbolic function, with clear saturation kynetics. In case of light, however, there is photoinhibiton for high light intensities. These type of responses are obviously taxa-specific and, for instance, diatoms are clearly adapted to higher nutrient concentrations than picoplanktonic cyanobacteria [21, 22, 27]. The competition for resources strongly determines where and when particular phytoplankton groups

flourish or fail. Yet, it is noteworthy that the competition outcome for light and nutrients depends not only on group specific resource uptake properties but also on other cell properties (e.g. motility, alternative nutritional strategies, and intra-cellular storage properties).

In addition to light and nutrients, the growth rate of phytoplankton is regulated by other environmental variables such as turbulence intensity, pH, temperature, salinity, oxygen concentration and the concentration of bioactive compounds, including cytotoxic compounds. Living phytoplankton often have specific environmental tolerances and, although some taxa exhibit an ubiquitous distribution, some phytoplankton species or assemblages show a spatially restricted distribution, sometimes with clear biogeographic provincialism. Latitudinal provincialism usually reflects aspects of tolerance to temperature and light whereas onshore to offshore provincialism is probably related to nutrient requirements or salinity tolerance. This is valid for all major phytoplankton groups including diatoms [6, 26], dinoflagellates [6, 29, 30], and coccolithophores [6, 41].

3.2. Regulation of phytoplankton loss rates

Phytoplankton loss rates in aquatic ecosystems were classically attributed to either grazing by multicellular organisms, particularly planktonic crustaceans, and passive sinking. The finding that phytoplankton assemblages are not dominated by diatoms and dinoflagellates but by pico- and nanoplanktonic naked forms, cells too small to be eaten directly by multicellular predators or to sink, clearly motivated the study of alternative phytoplankton predators [1]. Indeed, we now recognize that the main source of phytoplankton mortality is microzooplankton grazing. Microzooplankton is dominated by unicellular protists such as aplastidic nanoflagellates, ciliates, and heterotrophic dinoflagellates, and ingests, on average, 67% of phytoplankton daily production. This value varies only modestly among various habitats and regions [42]. The grazing impact of phagotrophic protists was firstly associated to small-sized phytoplankton cells. Yet, some protists, notably armored mixotrophic or heterotrophic dinoflagellates, use particular feeding mechanisms (e.g., tube feeding, pallium feeding) that allow ingestion of phytoplankton cells or colonies much larger than themselves [43]. Phagotrophic dinoflagellates often exhibit a high degree of prey selectivity; for example the mixotroph *Fragilidium* feeds exclusively on the genus *Ceratium*, the mixotrophs *Ceratium* and *Dinophysis* feed predominantly on ciliates, and the heterotroph *Protoperidinium* usually prefers diatoms. Therefore, grazing impact can also shape phytoplankton composition, favouring specific morphotypes or species, particularly grazing-resistant forms [44]. The fact that phytoplankton grazing is mainly undertaken by phagotrophic protists, and not by multicellular metazoans, has several consequences for the vertical flux of phytoplankton to deeper waters and its preservation in the geological record. When compared to metazoan grazers (e.g., copepods), phagotrophic protists clearly exhibit higher growth rates, almost as high as their phytoplanktonic prey. Consequently, phagotrophic protists are capable of rapidly responding to small increases in phytoplankton availability thus contributing to a relatively constant phytoplankton biomass, even during periods of increased phytoplankton production. This prompt response of phagotrophic protists decreases the likelihood of phytoplankton bloom occurrence as well as phytoplankton sinking fluxes, even in the case of large and high density phytoplankton taxa such as diatoms and coccolithophores. Since phagotrophic protists also exhibit lower sinking velocities, smaller-sized and lower density feacal pellets, minor vertical migrations, and higher metabolic activity and physiological losses (e.g., CO_2, dissolved organic and inorganic compounds) relative to metazoan grazers, the dominant effect of phagotrophic protists is to promote carbon retention in the water column minimizing vertical export and preservation in sediments [45, 46]. These effects are particularly enhanced under hydrographical conditions that promote phytoplankton assemblages dominated by small-sized cells [3, 47].

Although sinking, particularly in the case of large, non-motile and mineralized taxa, and predation are important loss routes for phytoplankton, other processes were recently integrated into the study of phytoplankton dynamics, namely viral induced lyses and regulated programmed cell death. Viruses are by far the most abundant biological entities in the oceans, comprising approximately 94% of the nucleic-acid-containing particles, and are responsible for the infection of both prokaryotic and eukaryotic phytoplankton groups. Since relationships between viruses and host populations are specific, viral infections can influence both host density and species composition, and are sometimes responsible for the termination of cyanobacteria, prasinophytes, pelagophytes, and raphidophytes phytoplankton blooms [48, 49]. The occurrence of diatom specific viruses was only recently referred [50]. The importance of eukaryotic parasites of phytoplankton (e.g., chytrids, Perkinsozoa, and parasitic dinoflagellates) was also recently discovered [51]. As a result of parasite induced lyses, living particulate organic matter (phytoplankton cells) is converted into dissolved organic compounds and cell fragments [49], thereby decreasing potential phytoplankton sinking rates and its accumulation into the sediment. Phytoplankton programmed cell death is an autocatalytic cell death induced by environmental stresses such as nutrient deprivation, intense light, excessive salt concentration or cell age. This process may be an important cause of mortality amongst diatoms, dinoflagellates, coccolithophores, and cyanobacteria. Cells that undergo programmed cell death exhibit a series of morphological and biochemical markers with specific associated genes and, as in case of viral lyses, the final products are again dissolved organic compounds and cell fragments [48]. Overall, death processes such as grazing by phagotrophic protists, viral lyses, lyses induced by eukaryotic parasites, and programmed cell death tend to promote the retention of carbon in the water column, thereby reducing its vertical export and preservation in the sediment.

Phytoplankton distribution in aquatic ecosystems is also regulated by more passive physical processes, related to vertical and horizontal advection of waters masses and specific circulation patterns. Dinoflagellate blooms, for instance, are sometimes related to physical accumulation [34]. Sinking, particularly important to non-motile, large, and mineralized phytoplankton cells or life-stages, should not be considered a purely passive process. In fact, phytoplankton sinking rates are clearly regulated by the physiological state of phytoplankton cells. Under particular physiological stresses, cell aggregation induced by increased sticking efficiency may further increase phytoplankton sinking rates [52].

3.3. Integration of bottom-up and top-down approaches

Phytoplankton dynamics was traditionally viewed as a result of resource-driven regulation, related to both light and nutrient availability. However, the relatively recent discovery of small-sized naked phytoplankton cells, phagotrophic protists, and planktonic viruses, currently considered key components of aquatic food webs, clearly raised our awareness to the importance of phytoplankton death in aquatic ecosystems [1, 48, 53]. The noteworthy changes in phytoplankton biomass and species composition induced by alterations of phytoplankton predators, described recently for different aquatic ecosystems, clearly indicate that loss processes are key controls underlying phytoplankton dynamics [2, 54]. In this context, it is important to emphasise that a phytoplankton bloom event can either arise from an increase in phytoplankton growth rate, a decrease in phytoplankton death rate, or both [55]. Hence, the comprehensive understanding of living phytoplankton dynamics should evidently be based on knowledge of both bottom-up (growth regulators) and top-down (biomass regulators) controls of phytoplankton. An interesting example that illustrates the need to contemplate both phytoplankton growth and mortality to comprehend dynamics of living phytoplankton, and eventually its traces in the geological record, are high nutrient low chlorophyll regions (HNLC). Indeed, low phytoplankton biomass in these oceanic regions can be attributable to entirely different factors. In the case of larger-sized phytoplankton cells, particularly diatoms, low biomass is typically attributed to a growth limitation by reduced iron bioavailability. Yet, for pico- and nanophytoplankton, low biomass mainly derives from high microzooplankton grazing rates relative to phytoplankton growth rates [56].

4. Links between dynamics of living phytoplankton and paleoenvironmental reconstruction

The distribution patterns of phytoplankton in present-day aquatic ecosystems are shaped by the interplay between processes that regulate cell growth and cell death. As mentioned in the previous section, cell growth and cell death are regulated by the internal environment of phytoplankton (e.g., specific environmental tolerances, resource uptake properties, cell size, density and morphology, alternative nutritional strategies such as mixotrophy or N_2 uptake, motility, intracellular storage capacities, grazing resistance properties), and by its external environment. The external environment includes variables dependent on the availability of resources (e.g., light intensity, concentration of CO_2 and dissolved inorganic macronutrients and micronutrients, availability of living prey in case of mixotrophs) and variables independent of resources (e.g., temperature, salinity, turbulence, ultraviolet radiation, bioactive compounds, activity of grazers, viruses, and eukaryotic parasites). All the environmental variables referred are integrated and reflected, in a simple or rather complex way, into several phytoplankton metrics. Indeed, the analysis of attributes of living phytoplankton, such as abundance, activity, and species composition, can be used to infer contemporaneous environmental conditions. Furthermore, this approach can also be extended into the reconstruction of paleoenvironmental and paleoecological conditions.

4.1. From phytoplankton composition and abundance to paleoenvironmental reconstruction

The occurrence of biogeographic provincialism, both in terms of near-shore to offshore and latitudinal gradients, in all major phytoplankton groups including diatoms [6, 26], dinoflagellates [6, 29, 30], and coccolithophores [6, 41], allows the use of key living phytoplankton species or assemblages as biological indicators of particular waters masses or environments. However, the geographic distribution and speciation of plankton, including phytoplankton, are also products of the geologic history of the oceans and continental barriers [57]. The recent trans-Arctic advection of the Pacific diatom *Neodenticula seminae* into the Atlantic induced by the melting of Arctic ice constitutes a notable example [58].

The use of fossil phytoplanktonic species or species associations as biozonal indices was frequently applied for reconstruction of paleoenvironments. In some cases, extinct taxa are also important tools in biostratigraphy [6, 26]. The paleoecological value of planktonic diatoms, in particular, is well established and particularly relevant in high-latitude ecosystems where calcareous microfossils tend to be sparse [6, 59, 60].

Relative abundances of particular functional groups of phytoplankton can also be used to unravel contemporaneous environmental conditions. For example, high relative abundances of cyanobacteria, particularly picoplanktonic forms or N_2 fixing forms, usually indicate oligotrophic ecosystems or oligotrophic phases. Indeed, cyanobacteria are particularly relevant in tropical oligotrophic ecosystems with low phytoplankton biomass and during warm stratified periods in temperate aquatic ecosystems [61]. In estuarine systems, for example, heatwaves and drought periods apparently promote cyanobacterial blooms [62]. On the contrary, high relative abundances of diatoms usually indicate eutrophic ecosystems or eutrophic phases. Indeed, diatoms are mostly relevant in eutrophic confined and exposed coastal ecosystems, particularly in upwelling regions, and during phytoplankton bloom events. These events usually take place during spring and autumn in mid-latitude ecosystems or during summer in high-latitude ecosystems and are typically associated to high resource levels and high turbulence regimes [21, 63]. However, anthropogenic eutrophication, sometimes aggravated by artificial freshwater regularization, may reduce the proportion of diatoms due to silica depletion [2]. In the case of dinoflagellates, high relative abundances of autotrophic dinoflagellates usually indicate low- to moderate resource levels. In fact, dinoflagellates are commonly abundant in low-turbulence low-nutrient stratified environments (e.g. highly stratified tropical ocean), during summer or autumn following diatom blooms or after upwelling relaxation [29]. Similarly to the dinoflagellates, high relative proportions of coccolithophores are usually associated to low- to moderate resource levels. Indeed, the majority of coccolithophores species occur in warmer stratified waters, particularly in sub-tropical gyres and upwelling areas [6, 37, 64]. Alterations in water quality evidently lead to alterations in phytoplankton species distribution or activity. For example, the recent increase of coccolithophore blooms in sub-Arctic areas was attributed to increased warmth and freshwater runoff [65]. Moreover, changes in the geographical extension of particular harmful dinoflagellate blooms as well as alterations in phytoplankton phenology were also related to climate changes in the North Atlantic [66, 67].

Relative abundances of particular species among each functional group of phytoplankton can also be used as indicators of contemporaneous environmental conditions and particular ecological niches. In the case of living diatoms, high proportions of pseudoplanktonic species (benthic pennate species observed in the water column) can trace resuspension events and thus be used as indicator of bentho-pelagic coupling [68], whereas elevated proportions of planktonic diatoms are usually used as a proxy of anthropogenic eutrophication [2]. As regards living dinoflagellates, high proportions of heterotrophic (e.g. *Protoperidinium*) to autotrophic species are usually indicative of a heterotrophic phase that commonly occurs after diatom blooms in more eutrophic ecosystems. Additionally, dinoflagellates with particular life-form properties explore "non-typical" habitat types, including shallow highly enriched ecosystems and physically-disturbed water masses [30]. Abundance and species/morphotype composition of living coccolithophores are also noticeably regulated by environmental conditions and particular species or species associations occupy specific ecological niches or hydrographic conditions [69, 70].

These relationships between phytoplankton metrics (species composition and abundance) and the external environment were also applied into paleoreconstruction studies. Indeed, fossil diatoms have been extensively used as proxies of past environmental parameters such as temperature, sea-level, water chemistry, pH, anthropogenic eutrophication, anoxia, habitat type, hydrographical conditions, productivity and biodiversity [6, 15, 16, 17, 19, 20, 59, 71, 72]. Dinocyst abundances, as well as heterotrophic to autotrophic dinocyst ratio, and relative abundances of specific dinocysts have been used as proxies of past environmental variables such as water temperature, salinity, nutrient input, cultural eutrophication, industrial pollution, hydrographical conditions (e.g., upwelling, river runoff, fronts), sea-level, sea ice-cover, and productivity in surface waters of aquatic environments [6, 12, 13, 18, 35, 73, 74, 75, 76]. Both abundance and composition of calcareous nannofossils have also been used as proxies of past environmental conditions (e.g., hydrographical conditions, climate changes) and ecological conditions [6, 77, 78, 79]. Since diatoms usually represent an opportunistic group indicative of phytoplankton blooms, fossil diatoms or its biomarkers are also used as proxies of nutrient availability and productivity [16].

Both biological oceanographers and paleoceanographers classically interpret phytoplankton dynamics as a result of resource-driven regulation, related to both light and nutrient availability. Hence, species-specific abundance or biomass values are generally used as direct indicators of environmental conditions. However, this general approach should be used cautiously. For example, high density of living phytoplankton or their fossil records do not necessarily imply elevated availability of inorganic nutrients. Indeed, cellular characteristics such as motility, mixotrophy, N_2 fixation or endosymbiosis with cyanobacteria (e.g., diatom *Hemialus* with endosymbiotic cyanobacteria *Richelia intracellularis*) can circumvent apparent resource limitation. Low contributions of particular phytoplankton groups may in fact indicate non-optimal growth conditions (e.g., light or nutrient limitation). However, they could also derive from high relative death rates, independently of absolute growth rates [3]. In the Ria Formosa coastal lagoon, for example, a high turbulence high nutrient ecosystem could, theoretically, favour diatoms. Yet, although diatoms represent the group with highest *in situ* replication rates, they represent only a modest fraction (3%) of phytoplankton abundance due to exceedingly high mortality rates [80]. Moreover, depleted fossil records can also be attributed to dense phytoplankton assemblages whose vertical fluxes to the sediment were restrained through protist grazing, viral lyses, programmed cell death, or reduced sinking rates. These processes could indeed explain the lack of correlation between cyanobacteria biomass in the water column and its specific fossil pigments in contemporaneously deposited sediments [81].

Interpretation of phytoplankton fossil record is further constrained by other problems (e.g., non-accurate recognition of species in the fossil record, species polymorphism and (pseudo)cryptic speciation, low preservation potential, selective degradation, lateral advection, extant lineages non-related to extinct lineages), not addressed within this study. However, the combination of multi-proxy approaches and other geochronology techniques will certainly permit more accurate interpretation of phytoplankton fossils in paleocological research, despite those methodological problems [10, 35, 82].

4.3. From phytoplankton elemental composition and species morphology to paleoenvironmental reconstruction

The external environment can also regulate other phytoplankton metrics, such as cell elemental composition and cell morphology. Nutrient availability, for instance, directly controls nutrient levels within the cells, while light intensity regulates the intracellular concentration of photosynthetic pigments as well as the proportion of specific photoprotective pigments or scytonemin in case of cyanobacteria [21, 23]. In diatoms, frustule morphology and elemental composition are controlled by silica and zinc availability, salinity, pollution and even grazing pressure [83, 84, 85, 86]. Regarding coccolithophores, the elemental strontium to calcium ratio in the coccolith fraction was found to covary with cell growth rate [62, 87]. Moreover, intra-specific variability in coccolith morphology and size can also be regulated by environmental conditions such as nutrient regime, temperate, salinity, and CO_2 concentration [88, 89, 90]. In dinoflagellates, environmental conditions such as water temperature, salinity, and light control both dinocyst production rates and dinocyst size and morphology [91, 92].

These relationships between environmental conditions and elemental or intra-specific morphological variability are also explored as paleoenvironmental proxies. The ratio between oxygen isotopes in the silica of fossil diatom frustules is used as a proxy of paleotemperature [6]. The proportions between photoprotective pigments and specific carotenoids or its geologically stable derivatives in diatoms and cyanobacteria have been used as indicators of past light intensity and ultraviolet radiation [10, 14]. In addition, coccolith sizes in particular species were used as proxies for paleotemperature and paleosalinity reconstruction [93, 94]. Sr/Ca ratio in coccoliths has been used as an indicator of coccolithophore productivity and, potentially, of coccolith vertical exportation rates [87, 95]. Indeed, morphological variability seems to hold a great potential for paleoenvironmental reconstruction [94]. However, it is important to note that both morphological features and elemental composition of specific phytoplankton groups are driven, simultaneously, by different environmental factors, including bottom-up and top-down forces. Therefore, their use as paleoenvironmental proxies should be undertaken cautiously.

4.4. From paleoenvironmental approaches to dynamics of living phytoplankton

The examples presented above clearly indicate that knowledge on the biology and ecology of living phytoplankton can be transposed into paleoenvironmental indicators. As stated previously, mechanisms that regulate phytoplankton growth as well as death should be considered both in paleoecological and in neoecological research. Additionally, the search for process oriented paleoecological indices (e.g., indicators of programmed cell death, predation, viral lyses), and the application of paleogenomic approaches could be of great potential for paleoenvironmental reconstruction.

Nevertheless, paleoenvironmental approaches are also extremely attractive tools to biologists, limnologists, and oceanographers. For instance, these approaches provide unique and critical information for the analysis of biological evolution (speciation and extinction) on Earth [6, 60]. The analysis of phytoplankton derived fossils in sediment-records also allows the extension of current observational time-series, both in space and time. In fact, the use of paleoecological approaches (e.g., morphological and chemical fossils) to characterize aquatic ecosystems under pristine ecological conditions was recently advocated in case of locations missing historical phytoplankton data sets [96]. Expanded time-series will certainly allow the test of relevant hypothesis over different time scales and a more precise analysis of natural and anthropogenically-driven environmental changes. Overall, this will undoubtedly improve our comprehensive understanding of the dynamics of living phytoplankton and ecosystem functioning.

5. References

[1] Sherr EB and Sherr BF 2000 *Microbial Ecology of the Oceans* ed DL Kirchman (New York: Wiley-Liss) pp 13-46

[2] Cloern JE 2001. *Mar. Ecol. Prog. Ser.* **210** 223-53

[3] Cloern JE and Dufford R 2005 *Mar. Ecol. Prog. Ser.* **285** 11-28

[4] Hays GC, Richardson AJ and Robinson C 2005 *Trends Ecol. Evol.* **20** 337-44

[5] Smetacek V and Cloern JE 2008 *Science* **319** 1346-48

[6] Armstrong HA and Braiser MD 2004 *Microfossils* (Oxford: Blackwell Publishing) p 29

[7] Falkowski PG, Katz ME, Knoll AH, Quigg A, Raven JA, Schofield O and Taylor FJR 2004 *Science* **305** 354-60

[8] Simoneit BRT 2004 *Advanc. Space Res.* **33** 1255-61

[9] Leavitt PR and Hodgson DA 2001 *Sedimentary Pigments Developments in Paleoenvironmental Research, Vol 3. Tracking environmental changes using lake sediments: terrestrial algal and siliceous indicators* ed Smol JO et al (Dordrecht: Kluwer Academic Publishers) p 295-325

[10] Verleyen E, Hodgson DA, Leavitt PR, Sabbe K and Vyverman W 2004 *Limnol. Oceanogr.* **49** 1528-39

[11] Zhao M, Mercer JL, Eglinton G, Higginson MJ and Huang C-Y 2006 *Org. Chemistry* **37** 72-97

[12] Dale B, Dale AL and Jansen JHF 2002 *Palaeogeogr. Palaeoclimatol. Palaeoecol.* **185** 309-338

[13] Radi T, Pospelova V, de Vernal A and Barrie JV 2007 *Mar. Micropaleontol.* **62** 269-297

[14] Leavitt PR, Cumming BF, Smol JP, Reasoner M, Peinitz R and Hodgson DA 2003 *Limnol. Oceanogr.* **48** 2062-69

[15] Rusak JA, Leavitt PR, McGowan S, Chen G, Olson O and Wunsam S 2004 *Limnol. Oceanogr.* **498** 1290-99

[16] Chmura GL, Santos A, Pospelova V, Spasojevic Z, Lam R and Latimer JS 2004 *Sci. Total Environ.* **320** 225-243

[17] Hodgson DA, Verleyen E, Squier AH, Keely BJ, Leng ML, Saunders KM and Vyverman W 2005 *Quat. Res.* **64** 83-99

[18] Dale B 2001 *Sci. Total Environ.* **264** 235-40

[19] Abrantes F 1991 *Mar. Micropaleontol.* **17** 285-310

[20] Nave S, Salgueiro E and Abrantes F 2003 *Mar. Geol.* **196** 21-35

[21] Miller CB 2004 *Biological Oceanography* (Oxford: Blackwell Publishing) p 402

[22] Timmermans KR, van der Wagt B, Veldhuis MJW, Maatman A and de Baar HJW 2005 *J. Sea Res.* **53** 109-20

[23] Groniger A, Sinha RP, Klisch M and Hader DP 2000 *J.Photochem.Photobiol.*B **58** 115-22

[24] Summons RE, Jahnke LL, Hope JM and Logan GA 1999 *Nature* **400** 554-57

[25] Coolen MJL, Talbot HM, Abbas BA, Ward C, Schouten S, Volkman JK and Damste JSS 2008 *Environm. Microbiol.* **10** 1783-03

[26] Guillard RRL and Kilham P 1977 *The Biology of Diatoms* ed D Werner (Oxford: Blackwell Scientific Publications) p 372-469
Willén E 1991 *Algological Studies* 62 69-106

[27] Sarthou G, Timmermans KR, Blain S and Tréguer P 2005 *J. Sea Res.* **53** 25-42.

[28] Coolen MJL, Volkman JK, Abbas B, Muyzer G, Schouten S and Damste JSS 2007 *Paleoceanography* **22** PA2211

[29] Taylor FJR and Pollingher U 1987 *The Biology of Dinoflagellates* ed FJR Taylor (Oxford: Blackwell Scientific Publications) p 398-533

[30] Smayda TJ and Reynolds CS 2001 *J. Plankton. Res.* **23** 447-61
Smayda TJ and Reynolds CS 2003 *J. Sea Res.* **49** 95-106

[31] Stoecker DK 1998 *Eur. J. Protistol.* **34** 281-90

[32] Skovgaard A 1996 *Phycologia* **35** 490-99

[33] Tang EPY 1995 *J. Plankton Res.* **17** 1325-35

[34] Persson A, Smith BC, Wikfors GH and Alix JH 2008 *Harmful Algae* **7** 798-801

[35] Rochon A, Eynaud F and de Vernal A 2008 *Mar. Micropaleontol.* **68** 1-5
[36] Throndsen J 1997 *Identifying marine phytoplankton* ed CR Thomas (San Diego: Academic Press) p 591-730
[37] Honjo S 1976 *Mar. Micropaleontol.* **1** 65-79
[38] Zondervan I 2007 *Deep-Sea Res.Part II* **54** 521-37
[39] Stoll HM and Ziveri P 2004 *Coccolithophores: from Molecular Processes to Global Impact* p 529-62
[40] Schippers P, Lurling M and Scheffer M 2004 *Ecol. Lett.* **7** 446-51
[41] McIntyre A and Bé AWH 1967 *Deep-Sea Res.* **14** 561-79
McIntyre A, Bé AWH and Roche MB 1970 *Transac.New York Acad.Sci.* **32** 720-31
[42] Calbet A and Landry M 2004 *Limnol. Oceanogr.* **49** 51-7
[43] Hansen PJ and Calado A 1999 *J. Eukaryot. Microbiol.* **46** 382-89
[44] Montagnes DJS, Barbosa AB, Boenigk J, Davidson D, Jürgens K, Macek M, Parry J, Roberts EC and Šimek K *in press Aquat. Microb. Ecol.* A1229
[45] Wassmann P 1998 *Hydrobiologia* **363** 29-57
[46] Strom S 2002 *Hydrobiologia* **480** 41-54
[47] Legendre L and Michaud J 1998 *Mar. Ecol. Prog. Ser.* **164** 1-11
[48] Bidle KD and Falkowski PG 2004 *Nature Reviews Microbiology* **2** 643-55
[49] Suttle CA 2007 *Nature Reviews Microbiology* **5** 801-12
[50] Tomaru Y, Shirai Y, Suzuki H, Nagumo T and Nagasaki K 2008 *Aquat. Microb. Ecol.* **50** 103-12
[51] Johansson M, Eiler A, Tranvik L and Bertilsson S 2006 *Aquat. Microb. Ecol.* **43** 289-302
[52] Kahl LA, Vardi A and Schofield O 2008 *Mar. Ecol. Prog. Ser.* **354** 3-19
[53] Kirchman DL 1999 *Nature* **398** 293-94
[54] Cloern JE, Jassby AD, Thompson JK and Hieb KA 2007 *Proc. Natl. Acad. Sci. U.S.A.* **104** 18562-5
[55] Riegman R, Kuipers BR, Noordeloos AAM and Witte HJ 1993 *Neth. J. Sea Res.* **31** 255-65
[56] Coale KH, Johnson KS, Fitzwater SE, Gordon RM, Tanner S, Chavez FP, Ferioli L, Sakamoto C, Rogers P, Millero F, Steinberg P, Nightingale P, Cooper D, Cochlan WP, Landry MR, Constantinou J, Rollwagen G, Trasvina A and Kudela R 1996 *Nature* **383** 495-501
[57] Zeitzschel B 1990 *Ecology of Marine Protozoa* ed GM Capriulo (New York: Oxford University Press) p 139-85
[58] Reid PC, Johns DG, Edwards M, Starr M, Poulin M and Snoeijs P 2007 *Global Change Biol.* **13** 1910-21
[59] Stoermer EF and Smol JP (eds) 2002 *The Diatoms: applications for the environmental and earth sciences* (Cambridge: Cambridge University Press) p 466
[60] Sims PA, Mann DG and Medlin LK 2006 *Phycologia* **45** 361-402
[61] Bell T and Kalff J 2001 *Limnol. Oceanogr.* **46** 1243-48
Weisse T 1993 *Advances in Microbial Ecology* vol 13, ed JG Jones (New York: Plenum Press) p 327-70
[62] Paerl HW, Valdes-Weaver LM, Joyner AR and Winkelmann V 2007 *Ecological Applications* **17** S88-S101
Johnk KD, Huisman JEF, Sharples J, Sommerijer B, Visser PM and Strooms JM 2008 *Global Change Biol.* **14** 495-12
[63] Longhurst A 1995 *Prog. Oceanogr.* **36** 77-167
Cébrian J and Valiela I 1999 *J. Plankton. Res.* **21** 429-44
[64] Gregg WW and Casey NW 2007 *Deep-Sea Res. Part II* **54** 447-77
[65] Smyth TJ, Tyrell T and Tarrant B 2004 *Geophys. Res. Lett.* **31** L11302 1-4
[66] Edwards M, Johns DG, Beaugrand G, Licandro P, John AWG and Stevens DP 2008 *SAHFOS Technical Report* **5** 1-8
[67] Edwards M and Richardson 2004 *Nature* **430** 881-84

[68] Facca C, Sfriso A and Socal G 2002 *Est. Coast. Shelf Sci.* **54** 773-92
[69] Schiebel R, Zeltner A, Treppke UF, Waniek JJ, Bollmann J, Rixen T and Hemleben C 2004 *Mar. Micropaleontol.* **51** 345-71
[70] Boeckel B and Baumann K-H 2008 *Mar. Micropaleontol.* **67** 255-73
[71] Olli K, Clarke A, Danielsson A, Aigars J, Conley DJ and Tamminen T 2008 *J. Mar. Syst.* **73** 284-99
[72] Bernárdez P, Prego R, Varela, M and Francés G 2008 *Cont. Shelf Res.* **28** 2496-508
[73] de Vernal A, Eynaud F, Henry M, Hillaire-Marcel C, Londeix L, Mangin S, Matthiessen J, Marret F, Radi T, Rochon A, Solignac S and Turon J-L 2005 *Quaternary Science Reviews* **24** 897-924
[74] Holzwarth U, Esper O and Zonneveld K 2007 *Mar. Micropaleontol.* **64** 91-119
[75] Ribeiro S and Amorim 2008 *Mar. Micropaleontol.* **68** 156-78
[76] Radi T and de Vernal A 2008 *Mar. Micropaleontol.* **68** 84-114
[77] Cachão M and Moita MT 2000 *Mar. Micropaleontol.* **39** 131-55
[78] Colmenero-Hidalgo E, Flores JA, Sierro FJ, Barcena MA, Lowemark L, Schonfeld J and Grimalt JO 2004 *Palaeogeogr. Palaeclimatol. Palaeecol.* **205** 317-36
[79] Su X and Wei KY 2005 Terrestrial Atmospheric and Oceanic Sciences **16** 95-111
[80] Barbosa AB 2006 Structure and dynamics of the microbial foodweb in the Ria Formosa PhD Thesys Universidade do Algarve p 517 (in portuguese)
[81] Bianchi TS, Ergelhaupt E, McKeee, BA, Miles S, Elmgren R, Hajdu S, Savage C and Baskaran M 2002 *Limnol. Oceanogr.* **47** 1537-44
[82] Young JR, Geisen M and Probert I 2005 *Micropaleontology* **51** 267-88
[83] Vrieling EG, Sun QY, Tian M, Kooyman PJ, Gieskes WWC, van Santen RA and Sommerdijk NAJM 2007 *Proc. Natl. Acad. Sci.U.S.A.* **104** 10441-46
[84] Ellwood MJ and Hunter KA 2000 *Limnol. Oceanogr.* **45** 1517-24
[85] Debenest T, Silvestre J, Coste M, Delmas F and Pinelli E 2008 *Aquat. Toxicol.* **88** 88-94
[86] Pondaven P, Gallinari M, Chollet S, Bucciarelli E, Sarthou G, Schultes S and Jean F 2007 *Protist* **158** 21-8
[87] Rickaby REM, bard E, Sonzogni C, Rostek F, Beaufort L, Barker S, Rees G and Schrag DP 2007 *Earth Planet. Sci. Lett.* **253** 83-95
[88] Batvik H, Heimdal BR, Fagerbakke KM and Green JC 1997 *Europ. J.Phycol.* **32** 155-65
[89] Green JC, Heimdal BR, Paasche E and Moate R 1998 *Phycologia* **37** 121-31
[90] Riesbesell U, Zondervan I, Rost B, Tortell PD, Zeebe RE and Morel FMM 2000 *Nature* **407** 364-66
[91] Zonneveld KAF and Susek E 2007 *Review of Paleobotany and Palynology* **145** 77-88
[92] Ellegaard M, Lewis J and Harding I 2002 *J. Phycol.* **38** 775-89
[93] Colmenero-Hidalgo E, Flores JA and Sierro FJ 20022 *Mar. Micropaleontol.* **46** 247-63
[94] Bollmann J and Herrle JO 2007 *Earth Planet. Sci. Lett.* **255** 273-88
[95] Stoll HM, Ziveri P, Shimizu N, Conte M and Theroux S 2007 *Deep-Sea Res.Part II* **54** 581-600
[96] Domingues RB, Barbosa AB and Galvão H 2008 *Mar. Pollut. Bul.* **56**: 1389-1395

IOP Conf. Series: Earth and Environmental Science **5** (2009) 012002 doi:10.1088/1755-1307/5/1/012002

Marine palynology and its use for studying nearshore environments

Anne de Vernal

GEOTOP, Université du Québec à Montréal, P.O. Box 8888, succursale "centre ville" Montréal, Qc, H3C 3P8 Canada

Email : devernal.anne@uqam.ca

Abstract. Palynology is the study of microfossils composed of highly resistant organic matter called palynomorphs. In the sediments of neritic environments, palynomorphs may include cysts of dinoflagellates, phycoma of prasinophytes, organic linings of benthic foraminifers and thecamoebians, in addition to inputs from the terrestrial vegetation (pollen grains and spores) or the freshwater biota (chlorococcales). Marine palynology is thus used for characterizing the type of sedimentary environment, identifying the source of organic matter in the sediment, and weighting the relative importance of fluvial and pelagic inputs.

Among marine palynomorphs, dinoflagellate cysts or dinocysts usually dominate the assemblages. Dinocysts comprise phototrophic and heterotrophic taxa and occur in almost all aquatic environments. Along the continental margins, assemblages are usually characterized by high species diversity and cyst concentrations reaching up to 10^5 cysts cm^{-3}. The distribution of dinocyst assemblages in sediments shows latitudinal patterns in addition to onshore to offshore gradients. Multivariate analyses illustrate close relationships between dinocyst assemblages and sea-surface parameters such as sea-ice cover, salinity, temperature, seasonality and productivity. Transfer functions developed from dinocysts permit the reconstruction of sea-surface temperature and salinity and the evaluation of past productivity, with applications dealing with climate changes and eutrophication. Dinocysts are also used for the study of harmful algal blooms since a few taxa relate to toxic species.

1. Introduction

The palynology is a science originally based on the observation of pollen grains and spores. Its early history dates from the end of the 17th Century with description of pollen grains made after observations under magnifying lenses (early microscope). The first pollen description was probably made by the English botanist Nehemia Grew. Palynology was developed as a research discipline from 1916 with the works of Lennart Von Post (1884 – 1951), a Swedish geologist who identified pollen grains at genera or species level in peat and established the first pollen diagrams to reconstruct vegetation and climate changes during the postglacial in northwest Europe.

The word « palynology » as such was introduced first in 1944 by the botanists Hyde and Williams from the Greek word *palunein* (= to sprinkle). In a strict sense, palynology is the study of pollen from plant flowers that is sprinkled as dust during the blooming season. In geology and paleobotany, the palynology has a much broader sense because the palynological slides prepared from sediments for the observation of pollen grains contain many other biogenic remains. By extention, palynology is the study of all organic-walled microfossils, also called palynomorphs (Table 1), which are observed on slides after treatments for pollen analysis performed in soft or indurated sediment [1]. These treatments notably consist in chemical digestion of carbonate and silicate particles with hydrochloric and hydrofluoric acids in order to concentrate resistant organic remains.

Palynomorphs in marine sediments include biological remains related to pelagic production such as cysts of dinoflagellates, phycoma of prasinophytes, lorica and spores of tintinnids, and copepod eggs. They also include remains of benthic organisms, notably organic linings of benthic foraminifers and thecamoebians. Marine sediments from neritic zones also contain allochthonous palynomorphs originating from the terrestrial vegetation and the freshwater biota, i.e., pollen grains of spermatophytes, spores of pteridophytes, bryophytes and fungi, organic walls of chlorococcales, etc. Palynomorphs may also include reworked microfossils resulting from the erosion of sedimentary outcrops. An exhaustive overview of palynology *sensu lato* and applications in biostratigraphy and paleoecolgy has been published by Jansonius and McGregor [2].

Table 1. List of microfossils recovered in sediments from continental margins (Palynomorphs in bold).

Microfossils and biological affinities	**Habitat of living organisms**					**Nature of the microfossil remains**			
	Terrestrial	Freshwater	Estuarine	Nearshore	Open Ocean	Organic	Siliceous	Carbonate	Agglutinated
ANIMAL									
Pteropods				XX	XXXX			XXXX	
Ostracods		XXXX	XXXX	XXXX	XXXX			XXXX	
Copepods (eggs)		**XXXX**	**XXXX**	**XXXX**		**XXXX**			
Sponge spicules		X	XX	XXXX	XXXX		XXXX	XX	
Chironomids (head capsules)		**XX**				**XXXX**			
Scolecodonts*			**X**	**XX**	**X**	**XXXX**			
PROTIST (heterotrophic)									
Benthic foraminifers			**XXXX**	**XXXX**	**XXXX**	**XX**	X	XXXX	XXXX
Planktonic foraminifers					XXXX			XXXX	
Thecamoebians	**XXXX**	**XXXX**	**XXXX**			**XXXX**			XXXX
Radiolarians				XX	XXXX		XXXX		
Dinocysts (Peridiniales, Gymnodiniales)		**XX**	**XXXX**	**XXXX**	**XX**	**XXXX**		X	
Tintinnids (spores and lorica)			**XXXX**	**XXXX**	**XXXX**	**XXXX**			XXXX
Chitinozoa*			**XX**	**XXXX**	**X**	**XXXX**			
PROTIST (mostly phototrophic)									
Silicoflagellates			XXXX	XXXX	XXXX		XXXX		
Diatoms		XXXX	XXXX	XXXX	XXXX		XXXX		
Coccolithophorids				XX	XXXX			XXXX	
Prasinophytes (Tasmanids, *Cymatiosphaera*)		**X**	**XXXX**	**XXXX**	**XXXX**	**XXXX**			
Dinocysts (Gonyaulacales)		**XX**	**XXXX**	**XXXX**	**XXX**	**XXXX**		X	
Chlorococcales (*Pediatrum*, *Botryococcus*)	**XXXX**	**XX**	**X**			**XXXX**			
Acritarchs*		**?**	**XX**	**XXXX**	**XX**	**XXXX**			
VEGETAL									
Pollen	**XXXX**	**XX**				**XXXX**			
Spores	**XXXX**	**XX**				**XXXX**			
Stomata	**X**					**XXXX**			
Phytolith	XXXX	X					XXXX		
FUNGI									
Spores and hyphae	**XXXX**					**XXXX**			

* Mostly Paleozoic XXXX = abundant; XXX = common; XX = occasional ; X = rare

Palynological approaches can be used for studying most sedimentary environments since the preservation of palynomorphs is usually good, because palynomorphs occur throughout the Phanerozoic, and because palynological assemblages include several proxies or tracers that can be analyzed simultaneously. In marine sediments of the continental margins, palynological assemblages are generally characterized by high concentration and high taxonomic diversity. They permit assessment on the source of organic matter and reconstruction of productivity and hydrographical conditions. Palynological assemblages are useful for various applications dealing with environmental issues such as productivity, eutrophication and climate changes. They led to application dealing with the biostratigraphy and paleogeography of the Paleozoic to recent. This manuscript presents an overview of the palynological content of sediments from neritic environment and a few examples of applications, with emphasis on palynological tracers of biogenic fluxes and dinocysts as proxy for the reconstruction of oceanographical parameters.

2. Palynological tracers of biogenic fluxes

2.1. Palynological tracers of terrestrial fluxes and hydrological inputs

Palynological tracers of terrestrial inputs are numerous. They are usually dominated by pollen grains and spores from vascular plants (spermatophytes and pteridophytes) in addition to spores from bryophytes. Pollen and spores derive from more or less long distance through atmospheric transport by winds, and hydrodynamic transport through runoff, rivers and marine currents. In general, the long distance atmospheric transport results in very low fluxes and concentrations in sediments [3]. Moreover, the assemblages are distorted, with an overrepresentation of anemophilous taxa relative to entomophilous taxa. Anemophilous taxa which have the embryo fertilized by pollen transported by wind (instead of insects) are usually characterized by a high pollen production, and by grain having a saccate form (i.e., equipped with sac-like appendices) favourable for long distance dispersal by wind transport. This is the case of *Pinus* pollen grains, which are characterized by bisaccate morphology and generally dominate assemblages at offshore locations and even in polar ice [4]. Pollen analyses along a nearshore to offshore transect in the Labrador Sea have shown that atmospheric transport is accompanied with an asymptotic decrease of pollen concentration from the coastline and an increase of relative proportion of *Pinus* [5]. Detailed pollen studies from sediment collected along European continental margins also led to demonstrate that most of the pollen in marine sediments is due to fluvial inputs from adjacent lands, therefore allowing direct comparison with terrestrial palynostratigraphy [6, 7]. Thus, as a first approximation, most pollen and spores in marine sediments can be associated to runoff and fluvial transport from adjacent land.

Other palynological tracers of fluvial transport include freshwater algae such as *Pediastrum* and *Botryococcus* that belong to the Chlorococcales (Table 1). A good example of the relationship between *Pediastrum* colonies and river discharge is provided by the analyses of surface sediment from the Arctic Ocean showing maximum concentrations in estuaries and deltas of the Beaufort, Laptev and Kara seas [8].

In nearshore sediments where runoff and fluvial inputs result in high organic carbon content, all sorts of palynological debris from terrestrial origin can be recovered. They include spores and hyphae of fungi, lignous debris, stomata, charcoal particles, etc.

2.2. Palynological tracers of pelagic fluxes and productivity

The cysts of dinoflagellates are amongst the most abundant pelagic microfossils in marine sediments. In offshore settings the calcareous forms are often abundant providing that calcium carbonate is well preserved. In epicontinental seas, estuaries and in nearshore environments, the organic-walled taxa also currently called dinocysts are usually abundant with fluxes ranging up to thousands cysts.cm^{-2}.yr^{-1} [5, 9]. The assemblages show a relatively high diversity of species and the occurrence of both phototrophic and heterotrophic species. Dinocyst thus relate to both primary phytoplanktonic productivity and zooplanktonic production. The concentration of dinocysts is, however, not a direct tracer of productivity since they represent a fragmentary picture of phytoplankton. Only 10-20 % of taxa yield organic walled microfossils [10]. Moreover, phototrophic dinoflagellates constitute only a part of the primary production, diatom and coccolithophorids having usually more rapid growth rates. Nevertheless, since heterotrophic dinoflagellates feed on primary producers and notably on diatoms, the abundance of their cysts indirectly provides indication of primary productivity. Heterotrophic dinocysts are thus particularly abundant in upwelling environments [11]. Although concentrations and fluxes of dinocysts do not permit quantitative paleoproductivity estimates, the distribution of assemblages (i.e., the percentages of species) show close relationship with the net primary production [12].

Among palynomorphs related to pelagic production, Prasinophytes such as *Tasmanites* and *Cymatiosphaera*, occasionally occur in modern nearshore environments, but have been recorded in high abundances mostly in Pliocene or older sediments [1, 13]. Other palynomorphs related to pelagic environments includes lorica and spores of tintinnids that belong to ciliates and occur occasionally in sediments [14]. Copepod eggs are also occasionally observed in palynological slides [15]. In general, the abundance and taxonomic diversity of palynomorphs related to planktonic production is higher along the continental margins than in open ocean environments.

2.3. Palynological tracers of benthic productivity

The shell of most foraminifers is composed of calcium carbonate or agglutinated material. Some taxa also possess shell with fossilisable organic lining in the inner part. The linings recovered in palynological preparations are often called microforaminifers. Organic linings of foraminifers characterize several calcareous and agglutinated benthic foraminifer taxa [16, 17]. *In vitro* dissolution of foraminifer shells led to identify some of the taxa that produce organic linings (cf. Table 2). These palynomorphs relate to a benthic production. Their occurrence in palynological slides often results from dissolution of foraminifer shells composed of $CaCO_3$. The ratio of organic linings to calcareous shells has been used as dissolution index [16].

Table 2. List of benthic foraminiferal taxa which were submitted to *in vitro* tests of dissolution of their shells in order to identify those yielding organic lining (X) [17].

Taxon name	Composition of the shell	Presence of organic lining
Adercotryma glomerata	Agglutinated	
Astrononion gallowayi	Carbonate	X
Bolivina pseudopunctata	Carbonate	X
Brizalina subaenariensis	Carbonate	X
Buccella hannai	Carbonate	
Bulimina aculeata	Carbonate	
Bulimina exilis	Carbonate	X
Bulimina marginata	Carbonate	X
Cassidulina laevigata	Carbonate	
Cassidulina reniforme	Carbonate	X
Cibicides lobatulus	Carbonate	X
Cribrostomoides crassimargo	Agglutinated	
Cribrostomoides jeffreysi	Agglutinated	
Elphidium excavatum	Carbonate	X
Elphidium incertum	Carbonate	X
Glandulina ovula	Carbonate	
Globobulimina auriculata	Carbonate	X
Gyroidinoides nitida	Carbonate	
Haplophragmoides sp.	Carbonate	
Hyperammina sp.	Agglutinated	
Islandiella helenae	Carbonate	X
Islandiella islandica	Carbonate	
Islandiella norcrossi	Carbonate	
Labradoricum sp.	Carbonate	
Lagena mollis	Carbonate	X
Lagena sp.	Carbonate	
Nonionella turgida	Carbonate	
Oridorsalis umbonatus	Carbonate	
Pseudoglandulina glanduliniformis	Carbonate	
Pullenia quinqueloba	Carbonate	
Pyrgo williamsoni	Carbonate	
Quinqueloculina seminulum	Carbonate	
Quinqueloculina stalkeri	Carbonate	
Reophax scottii	Agglutinated	X
Robertinoides charlottensis	Carbonate	
Sahulia conica	Carbonate	
Spiroplectammina biformis	Agglutinated	X
Textularia earlandi	Agglutinated	
Triloculina sp.	Carbonate	

3. Using palynological tracers for the characterization of sedimentary environments

3.1. Tracing the origin of organic matter

Among palynomorphs, the most common terrestrial inputs are pollen grains whereas dinocysts usually dominate fluxes of pelagic origin. On this basis, the pollen vs. dinocyst ratio can be used as index of the terrestrial component in sediment. Example is provided by the analyses that were performed in the Laurentian channel throughout the St. Lawrence Estuary-Gulf estuarine system [18]. The palynological analyses show that pollen/dinocyst ratio decreases from land to ocean, parallel to an increase of $\delta^{13}C$ in organic matter (Figure 1). Both proxies show an upstream-downstream gradient that relates to the decrease of terrestrial contribution relative to marine fluxes towards the end member of the estuarine system. Pollen/dinocyst ratio can thus be used as terrestrial vs. marine input or "continentality" index.

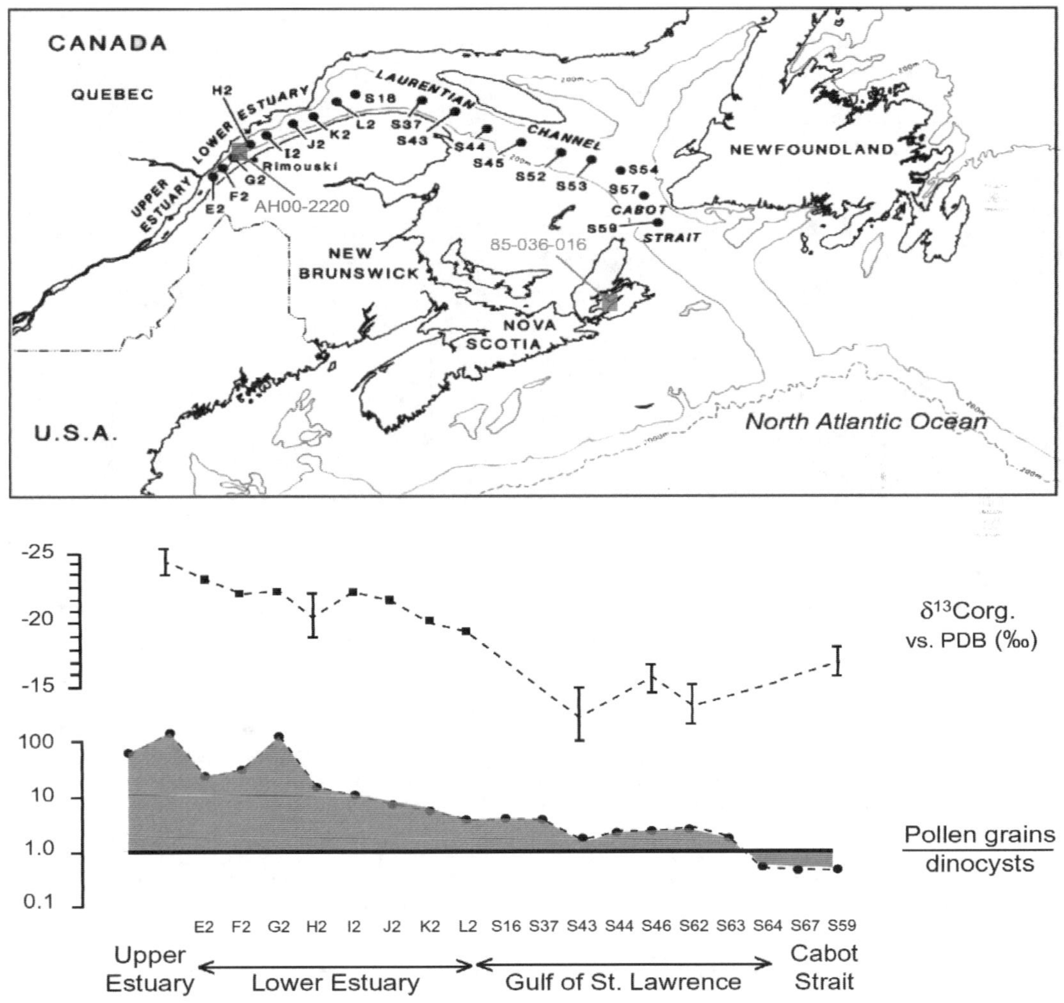

Figure 1 - Location map of surface sediment samples collected along an upstream-downstream transect from the St. Lawrence River to the North Atlantic Ocean (black dots in the upper panel), $\delta^{13}C$ of organic matter and pollen/dinocyst ratio (lower panel), which provides indication on the source (terrestrial vs. marine) of the organic content [18].

3.2. Reconstructing sea-level changes

Based on the respective occurrence of terrestrial and marine palynomorphs, it is possible to assess on the freshwater vs. marine character of sediments, thus to reconstruct changes in sea-level. One example is presented in Figure 2. It is from the Lake Bras d'Or, which is a marine water body of Nova Scotia, eastern Canada. The lake is connected to the Atlantic Ocean by a sill about 8 meters deep. The analyses of sediments from core 85-036-016 show an important transition in palynological assemblages and isotopic composition of organic matter at 1.6 meters. Below 1.6 m, the palynological assemblages relate to freshwater biota with abundant spores of aquatic plants, chlorococcale remains and freshwater dinoflagellate cysts. Above 1.6 m, freshwater taxa disappear and the assemblages are dominated by marine dinocysts [19]. The freshwater to marine transition identified from palynological assemblages coincides with an important shift in $\delta^{13}C$ of organic matter, which also indicate relative increase of marine pelagic production relative to terrestrial input [20]. The date of the transition is about 4000 years BP. Therefore, assuming the sill depth remained unchanged, one may calculate mean relative sea-level rise of about 20 cm per century [19, 20]. In the context of Eastern Canadian margins, the relative sea-level variation is linked with the postglacial isostatic rebound at the periphery of the Laurentide Ice Sheet that reached maximum extension about 21 000 years ago.

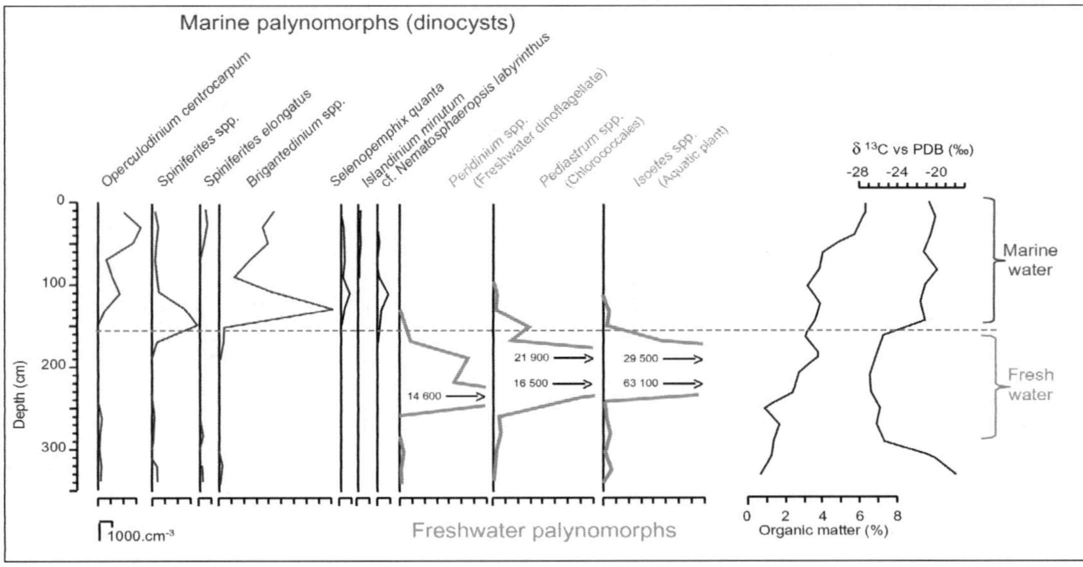

Figure 2 - Palynological assemblages [19], and organic carbon content and isotopic composition [20] of core 85-036-016P collected in the Lake Bras d'Or, Cape Breton Island, Nova Scotia (for site location, see blue square in Figure 1). The basin is connected to the sea through a channel having its sill at 8 meters below present sea-level. The freshwater to marine assemblage transition occurred at about 4000 years BP, which indicate local sea-level rise of about 0.2 cm per year.

3.3 Identifying floods

Palynological analyses permit the identification of changes in sedimentary regime and dramatic sedimentary events, such as flood and debris flow. In general, flood event would be characterized by reworked material from terrestrial origin and limited pelagic inputs. Such is the case, for example, of the flood layer associated with the 1663 earthquake that is observed in the Saguenay fjord, adjacent to the St. Lawrence Estuary [21].

3.4. Evaluating calcium carbonate preservation

Some benthic foraminifer species produce organic linings (OL) that fossilize as palynomorphs and could be used as proxy of both benthic production and calcium carbonate preservation (see section

2.3). The ratio of organic linings to calcareous shells (OL/CS) has been used to assess on $CaCO_3$ dissolution in estuarine and deep sea environments [16, 21-22]. For example, the OL/CS ratio provided evidence of deepening of the lysocline depth at the transition from the last glacial to interglacial in the Gulf of Alaska, subpolar North Pacific [22]. As another example, close to infinite OL/CS ratio in Holocene sediment of subarctic Baffin Bay relates the establishment of shallow calcium compensation depth during the present interglacial [16].

The OL/CS ratio is a proxy for $CaCO_3$ preservation that may vary with large scale changes of ocean circulation and chemical properties of sea water. However, synsedimentary and postsedimentary diagenetic processes also play an important role. Moreover, sampling artefact cannot be discarded, especially in the case of organic rich sediments. When opening the core, the availability of O_2 can result in oxidation of organic matter with production of CO_2, which thus lower the pH and affect the $CaCO_3$ preservation. Similarly, organic rich sediment often contains FeS, which produces sulphuric acid at the contact with the air, thus lowering the pH and dissolving carbonate. Reproducibility tests have demonstrated that sampling procedures and sediment storage may differentially affect the preservation of calcium carbonate [17].

4. Dinocysts as proxy of hydrographical conditions and productivity

4.1 Biology of dinoflagellates and their cysts

Dinoflagellates are microscopic unicellular organisms belonging to the division of Dinoflagellata [23]. Their populations develop in most types of aquatic environments, from lakes to open ocean, and from equatorial to arctic settings. Dinoflagellates have various trophic behaviour, some being phototrophic, other being heterotrophic or mixotrophic, while some are symbionts (zooxanthellae) of marine invertebrates such as corals for example. When blooming, dinoflagellates can cause red tides. Because a few species produce neurotoxines that may be concentrated by filtering organisms such as shellfish, red tides are an issue for fisheries.

Most dinoflagellates have a complex life cycle involving several stages, asexual and sexual, motile and non-motile. During the course of the sexual reproduction, many dinoflagellate form a diploid cell protected within a cyst, which permit survival of the organism during a dormancy period of variable length [23]. The cyst of many species is composed of highly resistant organic matter, similar in composition to the sporopollenin of pollen grains. The organic-walled cysts also called dinocysts were previously known as "hystrichospheres" in the paleontology literature.

Organic walled dinoflagellate cysts or dinocysts represent dormancy stage in relation with sexual reproduction. They provide a fragmentary picture of original populations.

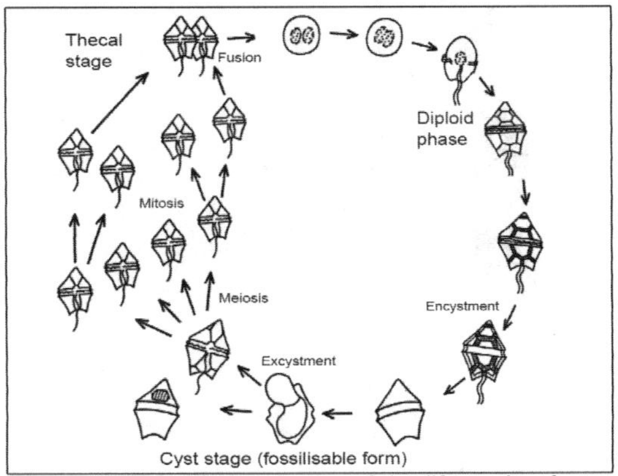

Figure 3. Simplified scheme of the life cycle of dinoflagellates. Note that only the cyst can be fossilized

4.2 Dinocyst distribution in relation with environmental parameters

Fossil dinocysts are mostly known from marine sediments, and appear to be particularly abundant along continental margins, in estuaries, epicontinental seas, continental shelves and slopes. Dinocysts were widely used in biostratigraphy and paleoecology of the Mesozoïc and Tertiary. In the fields of paleoecology and environmental sciences, the study of dinocysts is of growing interest. Because they are very resistant, dinocysts are generally well preserved in sediment despite dissolution that may affect calcareous or siliceous biological remains. Moreover, the development of reference databases from the analysis of surface sediment samples [12, 24-27] permitted to document close relationships between the distribution of dinocyst assemblages and sea-surface parameters.

Figure 4 - Northern Hemisphere dinocyst database established after standardization of laboratory procedures and taxonomy [e.g., 12, 24, 25]. The dashed circle corresponds to the polar circle and the grey zone to perennial sea-ice. Of the 64 taxa included into the database, 57 are recorded in the North Pacific samples (green dots), 56 in the North Atlantic (red triangles) and 37 in the Arctic (blue squares).

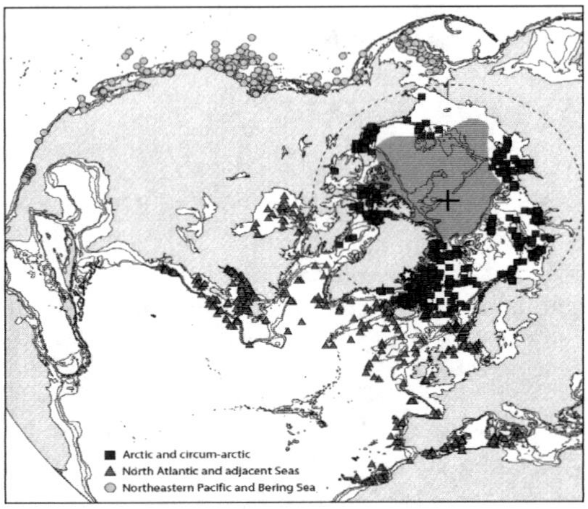

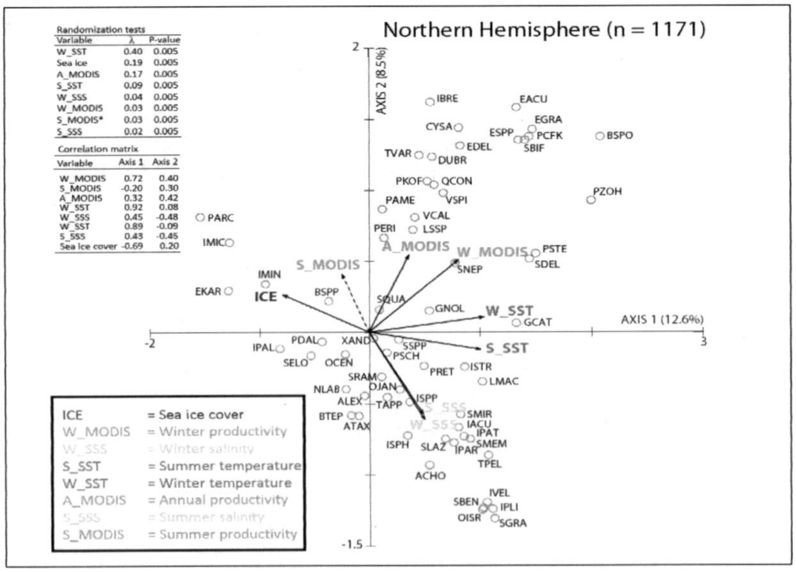

Figure 5 - Results of canonical correspondence analyses showing the ordination of species and that of oceanographical parameters. Data and figure are from Radi and de Vernal [12].

For several years efforts have been made for standardization of laboratory procedures and taxonomical nomenclature in order to establish a dinocyst assemblage database for quantitative data treatments. The standardized Northern Hemisphere database includes a total of 64 taxa (see database on the GEOTOP website) and close to 1200 reference sites allowing multivariate analysis. Canonical correspondence analyses were performed on assemblages (the percentages of the 64 taxa) and oceanographic parameters: (seasonal and annual productivity, winter and summer temperature and salinity, sea-ice cover). The results demonstrated that temperature exerts a primary control on the distribution of assemblages. They also show that all the above mentioned parameters are determinant [12, 25-27] and that they are not inter-correlated, except the summer and winter salinity (see Figure 5).

The analyses performed on regional databases, i.e., North Pacific, Arctic and North Atlantic illustrate distinct weighting of taxa depending upon the region analyses [12, 28], suggesting some regionalism in the distribution of taxa and their relations to hydrographic conditions. For example, the eastern North Pacific dinocyst distribution appears to be first determined by productivity, which seems logical since the oceanography is largely controlled by upwelling intensity [12]. As another example, the salinity is most determinant in the northwest domain of the North Atlantic, which is also consistent taking into account the large salinity gradient that are observed [28].

The World distribution of dinocysts is known from various studies which were summarized by Marret and Zonneveld [25]. It shows distinct assemblages in mid-high latitudes of Northern and Southern Hemispheres [27, 29] and suggests evolution and extinction of species that differ in high latitudes of the Northern and Southern Hemisphere. In view of the general decline in the number of dinocyst species during the late Cenozoic in the Northern North Atlantic [30], one may hypothesize a relation between the regional appearance-disappearence of dinocyst taxa and the large amplitude climate changes at the onset of continental glaciations.

4.3 Transfer functions applied to dinocyst assemblages for paleoceanographical reconstruction

Various transfer functions techniques [31] have been tested with dinocyst databases. To date, the modern analogue technique (MAT) seems to be the most appropriate for quantitative reconstruction of past oceanographical conditions. MAT is based on the similarity between fossil and modern spectra and assumes that fossil assemblages developed in environmental conditions similar to their modern analogues. Unlike calibration approaches such as the artificial neural network and the Imbrie and Kipp technique, MAT does not use a defined relationship between assemblages and environmental parameters. In this sense, MAT is not susceptible of extrapolation and should give robust results inasmuch as good analogues exist.

Most studies using paleoceanographical reconstructions based on dinocysts followed the same procedures [26], which include the log transformation of taxa occurrence prior to search for analogues in the reference database. Such a transformation is necessary to emphasize the weight of accompanying taxa, which often have narrow ecological affinity whereas the most abundant taxa are often ubiquitous. MAT using the R software [http://www.r-project.org] have been adapted by Guiot from PPPbase [32; http://www.imep-cnrs.com/pages/3pbase.htm].

Table 3. Accuracy of past sea-surface condition estimates based on MAT applied to dinocyst assemblages. The results are from the Northern Hemisphere database (n = 1171 sites and 64 taxa; data available in paleoceanographic data pages under the GEOTOP webpage (www.GEOTOP.ca; cf. also references 12 and 25)). "r^2" refers to the coefficient of correlation between observed and estimated values and the RMSE is the error in the reconstruction, which corresponds to the standard deviation (1σ) of the difference between observed and estimated values.

Parameter	Standard deviation (1σ) Instrumental data	r^2	RMSE
Winter temperature	±1.08 °C	0.97	1.19 °C
Summer temperature	± 1.55 °C	0.95	1.71 °C
Summer salinity (>30)	± 0.63	0.9	0.62
Sea-ice cover	± 1.19 months/yr	0.89	1.24 months/yr
Winter productivity	na	0.92	11.5 gC/m^2 (~ 20%)
Summer productivity	na	0.79	23 gC/m^2 (~ 25%)
Annual productivity	na	0.84	55 gC/m^2 (~ 18%)

Validation tests are made to evaluate the performance of the approach and to calculate the error of prediction (see Table 3). The reliability of the approaches is given by the coefficient of correlation (r^2) between observed and estimated values, whereas the accuracy is provided by the root mean square error (RMSE) that corresponds to the standard deviation of the difference between observed and estimated value. In the case of the Northern Hemisphere database, the error of the reconstruction is close to the standard deviation of instrumental values.

The application of MAT to dinocyst assemblages has permitted reconstruction of hydrographic conditions (temperature, salinity, sea ice) for times series of the Pleistocene and Holocene, mostly in middle to high latitudes of the Northern Hemisphere [26], but also in the northern Pacific [26] and southern Ocean [28]. It has also been used for mapping oceanographical conditions (salinity, temperature, sea-ice in the northern North Atlantic Ocean during the last glacial maximum, about 21 000 years ago [26].

5. Examples of applications of marine palynology and dinocyst studies

In neritic environments, applications of dinocyst studies include the evaluation of productivity changes in relationship with recent eutrophication, and the reconstruction of sea-surface temperature and salinity conditions. Applications also include the study of harmful algal blooms since a few dinocyst taxa relate to toxic species. Two examples are presented below, one from the estuarine system of the St. Lawrence River, eastern Canada. The other is from the Pacific coast of Mexico.

5.1. Eutrophication in the Estuary of St. Lawrence

In the Estuary of St. Lawrence, decreasing concentration of dissolved oxygen in bottom water is a source of concerns [32]. It has been proposed that the recent lowering of dissolved oxygen is related to increased primary productivity and consumption of oxygen through oxidation of organic matter. In order to investigate the cause of recent hypoxia and to verify the hypothesis of recent eutrophication and increase of carbon fluxes in the lower Estuary of St. Lawrence, palynological and geochemical analyses were carried out on box core sediments [33]. Results show an increase in the accumulation rate of dinocysts over the last decades together with an increase in organic carbon content and a shift in the isotopic composition of organic carbon, strongly suggesting increased primary productivity and organic carbon fluxes to the sea floor (Figure 6). The application of MAT to dinocyst assemblages also suggests increased production after 1970 AD. In this case, the data corroborate that increased productivity in the Estuary of St. Lawrence may have played a role in the eutrophication and hypoxia starting after 1970 AD. However, the data also show large amplitude variations through time, which suggest important variation of productivity due to natural forcing.

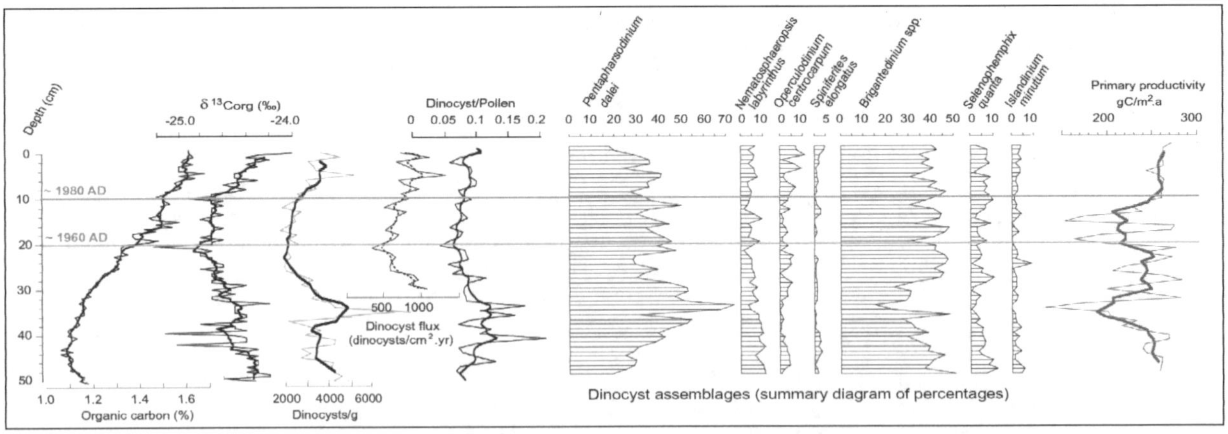

Figure 6 - Organic carbon content, isotopic composition of carbon, palynological assemblages [33] and estimates of primary productivity [12] from core AH00-2220 collected in the St. Lawrence Estuary (see red square in Figure 1 for core location).

5.2. Harmful algal bloom along the Mexican coasts

Harmful algal blooms (HABs) and red tides are a source a concern, especially in coastal areas where the marine resources are economically important such as those of Mexico [34]. Many taxa responsible for HABs are dinoflagellates and a few of them are known to produce cysts, which could generate blooms decades after their settling. Thus, dinocysts in sediments deserve special attention since they may help identifying areas susceptible to HABs and permit investigation of the HAB history beyond human observations. As an example, the study performed in a core collected in the Gulf of Tehuantepec along the western Mexican margin has shown abundant dinocysts with fluxes ranging up to 500 cysts cm^{-2} yr^{-1} and continuous occurrence of the toxic HAB species *Polyspaeridium zoharyi* (the cyst of *Pyrodinium bahamense*) over the last two centuries [35].

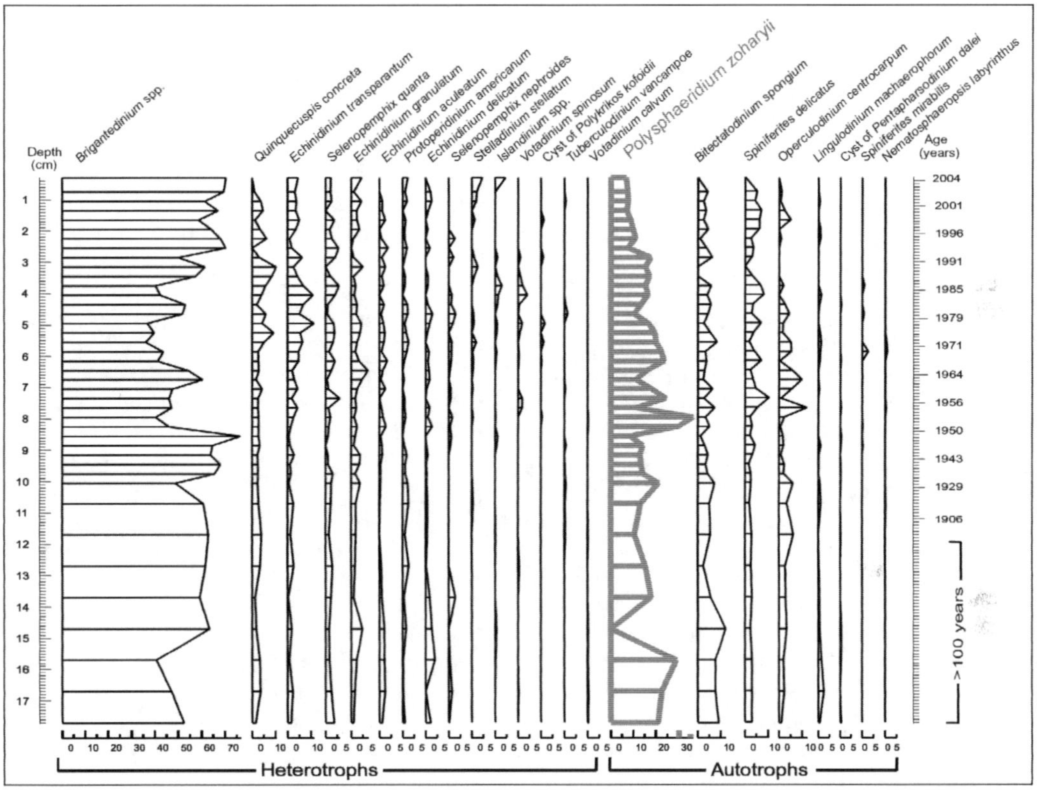

Figure 7 - Diagram of dinocyst taxa percentages vs. depth in a core from the Gulf of Tehuantepec off western Mexico. *Polyspaeridium zoharyi* (in orange) is the cyst of *Pyrodinium bahamense*, which is a toxic HAB species [36].

6. Conclusion

Almost all biota provide organic remains preserved in sediment as palynomorphs, which can be used as tracers of the origin of biogenic fluxes and permit characterizing past depositional environments. This is particularly useful in littoral, shelf and estuarine settings, where changes in hydrological regime, sea-level, hydrographical conditions in addition to the anthropogenic activity may affect the sedimentary dynamics and productivity in pelagic and benthic environments.

Marine palynology therefore allows many applications in environmental sciences and geosciences. In particular, dinocysts provide information on hydrographical changes, productivity variations, and red tides in the past, which are important issues to address for evaluating the natural variability of the coastal environments and distinguishing the impact of anthropogenic forcing.

7. Reference

[1] Traverse A 1988 *Paleopalynology* Unwin Hyman, 600 p.
[2] Jansonius J and McGregor DC, Eds. 1996 *Palynology: principles and applications* American Association of Stratigraphic Palynologist Foundation, College Station TX
[3] Heusser LE and Balsam WL 1977 *Quat. Res.* **7** 45
[4] Bourgeois JC 1990 *J. Glaciology* **36** 340
[5] Rochon A and de Vernal A 1994 *Can. J. Earth Sci.* **31** 15
[6] Sánchez Goñi MF, Eynaud F, Turon JL and Shackleton NJ 1999 *Earth Planet. Sci. Lett.* **171** 123
[7] Naughton F, Sanchez Goñi MF, Desprat S, Turon JL, Duprat J, Malaizé B, Joli C, Cortijo E, Drago T and Freitas MC 2007 *Mar. Micropal.* **62** 91
[8] Matthiessen J, Kunz-Pirrung M and Mudie PJ 2000 *Intern. J. Earth Sci.* **89** 470
[9] de Vernal A, Rochon A, Turon J-L, Matthiessen J1997 *GEOBIOS* **30** 905
[10] Head MJ 1996 In: Jansonius J and McGregor DC 1996 *Palynology: principles and applications* American Association of Stratigraphic Palynologist Foundation, College Station TX 1197.
[11] Gaines G and Elbrachter M 1987 In: Taylor FJR *The Biology of Dinoflagellates* Blackwell Scientific, Oxford 224
[12] Radi T and de Vernal A 2008 *Mar. Micropal.* **68** 84
[13] de Vernal A and Mudie PJ 1989 *Proc. Ocean Drilling Program* **105B** 401
[14] Lipps, JH Ed 1993 *Fossil prokaryotes and protists* Blackwell Scientific Publications
[15] Van Waveren IM 1994 *Scripta Geologica* **105** 67
[16] de Vernal A, Bilodeau G, Hillaire-Marcel C and Kassou N 1992 *Geology* **20** 527
[17] Leduc J 2001 *Study of benthic foraminiferal populations in the sediments of the Saguenay Fjord.* MSc Thesis, UQAM, Montréal Qc, 63 p.
[18] de Vernal A, Giroux L and Hillaire-Marcel C 1991 *Palynosciences* **1** 145
[19] de Vernal A and Jetté H 1987 *Geological Survey of Canada* **87-1A**, 11
[20] Hillaire-Marcel C 1987 *Geological Survey of Canada* **87-1A**, 89
[21] St-Onge G, Leduc J, Bilodeau G, de Vernal A, Devillers R, Hillaire-Marcel C, Loucheur V, Marmen S, Mucci A and Zhang D 1999 *Géogr. Phys. Quat.* **53** 339
[22] de Vernal A and Pedersen T 1997 *Paleoceanography* **12** 821
[23] Fensome RA, Taylor FJR, Norris G, Sarjeant WAS, Wharton DI and Williams GL 1993 *Micropaleontology Special Publication* **7** 351 p.
[24] Rochon A, de Vernal A, Turon J-L, Matthiessen J and Head MJ 1999 *Special Contribution Series* **35** American Association of Stratigraphic Palynologists, Dallas TX
[25] Marret F and Zonneveld KAF 2003 *Rev. Palaeobot. Palynol.* **125** 1
[26] de Vernal A, Eynaud F, Henry M, Hillaire-Marcel C, Londeix L, Mangin S, Matthiessen J, Marret F, Radi T, Rochon A, Solignac S and Turon J-L 2005 *Quat. Sci. Rev.* **24** 897
[27] de Vernal A and Marret F 2007 In Hillaire-Marcel C and de Vernal A *Proxies in Late Cenozoic Paleoceanography* Elsevier 371
[28] de Vernal A and Radi T 2008 *Proceedings of the Eighth International Conference on Modern and Fossil Dinoflagellates* Montreal
[29] Marret F, de Vernal A, Benderra F and Harland R 2001 *J. Quat. Sci.* **16** 739
[30] de Vernal A and Mudie PJ 1989 *Proceedings of the Ocean Drilling Program* **105B** 387
[31] Guiot J and de Vernal A 2007 In Hillaire-Marcel C and de Vernal A *Proxies inLate Cenozoic Paleoceanography* Elsevier 523
[32] Gilbert D, Sundby B, Gobeil C, Mucci A and Tremblay G-H 2005 *Limnol. Oceanogr.* **50** 1654
[33] Thibodeau B, de Vernal A and Mucci A 2006 *Marine Geol.* **231** 37
[34] Hallegraeff CM, Anderson DM, Cembella AD eds. 2003 *Monographs on Oceanographic Methodology* **11** Paris, UNESCO.
[35] Vásquez-Bedoya LF, Radi T, Ruiz-Fernández AC, de Vernal A, Machain-Castillo ML, Kielt J-F and Hillaire-Marcel C 2008 *Mar. Micropal.* **68** 49

Acknowledgments
This manuscript uses the published and unpublished results of several years of work in marine palynology undertaken in collaboration with many students and colleagues. Special thanks are due to

Maryse Henry for her help as lab manager of the palynological laboratory at GEOTOP-UQAM. Most of the financial support was provided by the Natural Sciences and Engineering Research Council (NSERC) of Canada and the *Fonds Québecois de Recherche sur la Nature et les Technologies* (FQRNT).

The ecology and biological affinity of Arctic dinoflagellates and their paleoceanographical significance in the Canadian High Arctic

A Rochon

ISMER-UQAR, 310 allée des Ursulines, Rimouski QC, Canada G5L 3A1

Andre_rochon@uqar.qc.ca

Abstract. Dinoflagellates are eukaryotic organisms and constitute an important group of marine primary producers. Approximately 10-15% of living dinoflagellates produce a highly resistant dormant cyst that is fossilisable, and which constitute an excellent proxy indicator of the upper water column conditions and productivity. Relatively little is known on the distribution in time and space of the dinoflagellate life cycle (i.e., vegetative and cyst stages) in the Canadian Arctic; most studies usually focusing on other groups of organisms (e.g., diatoms). Here, we present information on the ecology of dinoflagellate cysts and how they relate to their counterpart plankton stages. We discuss the importance of considering the biological affinities of dinoflagellates cysts and their relevance for paleoceanographical interpretations. We also provide insight on the actual lack of such knowledge for the Canadian Arctic cyst and plankton assemblages. **Keywords: Dinoflagellate, dinoflagellate cyst, cyst-theca-relationships, Arctic, paleoenvironments**

1. Introduction

Dinoflagellates are single-celled eukaryotic organisms that inhabit both freshwater and marine environments from equatorial to polar latitudes. In marine environments, they constitute together with diatoms and coccolithophores an important group of primary producers. About half of the extant dinoflagellates are autotrophic, or mixotrophic (combination of phototrophy and heterotrophy), while the other half are heterotrophic, and some are parasitic. Approximately 10-15% of dinoflagellates form a cyst composed of a highly resistant polymer, as part of their life cycle [1]. The cyst stage, or hypnozygote, is believed to be a benthic phase in the life cycle of the dinoflagellates [2] and can remain dormant in the sediments from several days to years. Quaternary dinoflagellate cysts, known as dinocysts, assemblages are widely used by marine palynologists as a proxy indicator of conditions in the upper part of the water column [3, 4, 5], eutrophication [6, 7, 8] and productivity [9, 10]. Over the last few decades, the distribution of dinocyst assemblages in the surface sediments of the ocean basins, especially the Arctic basins, have been the subject of numerous studies in an effort to develop their potential for paleoceanographical reconstructions [11, 12, 13, 14, 15, 16, 17]. So far, approximately 31 dinocyst taxa have been observed in surface sediments of the Canadian Arctic [3, 14, 41, 45]. Several studies reported the distribution of the vegetative motile forms of dinoflagellates in the Arctic [18, 19, 20, 21, 22, 23, 24, 25, 26, 27] and ~192 taxa belonging to 42 genera have been recorded from the water column of the Canadian High Arctic [18, 23, 24, 27, 28, 29]. However, our knowledge of the relationships between the plankton motile forms and their cyst counterpart in the sediments is rather limited. For example, the cyst stage of several motile forms collected from the phytoplankton have not been recorded in surface sediments yet, while several motile forms related to the known Arctic cyst taxa have not been observed in Arctic phytoplankton samples or described yet. The knowledge of the biological affinity or cyst-theca relationships for Arctic dinoflagellates is important because it provides useful information on dinoflagellate life cycle. In addition, the increasing understanding of the motile form will prove extremely useful for paleoceanographical reconstructions and interpretation. A better knowledge of the distribution of the dinoflagellate motile forms will also provide hints on the transport mechanisms affecting the dinocysts after their formation and subsequent deposition in the surface sediments.

2. Dinoflagellate ecology and cyst-theca relationships

Dinoflagellates have a relatively complex life cycle that includes a motile plankton or vegetative stage, and some include a resting or cyst stage (Figure 1). The motile forms are characterized by the presence of two flagella, one longitudinal and the other located in a depression or cingulum, around the cell. The cell is comprised within a membrane, the amphiesma, which can include a series of cellulosic plates (referred to as "armored" or thecate), while others lack those cellulosic plates and are referred to as "naked" or "athecate". Dinocysts are either composed of dinosporin, a complex biomacromolecular substance composed of phenolic, alcoholic and/or carboxylic hydroxides, and fatty acids ranging from C14 to C16, accompanied by tocopherols and sterols (mostly cholesterol and dinosterol) [30], or calcium carbonate. The morphology of most motile and cyst species are often very different, which makes it difficult to understand the complete life cycle. Also, the motile and cyst forms were studied separately by biologists and marine palynologists respectively. This gave rise to two distinct nomenclature systems, which further complicates the understanding of their life cycle.

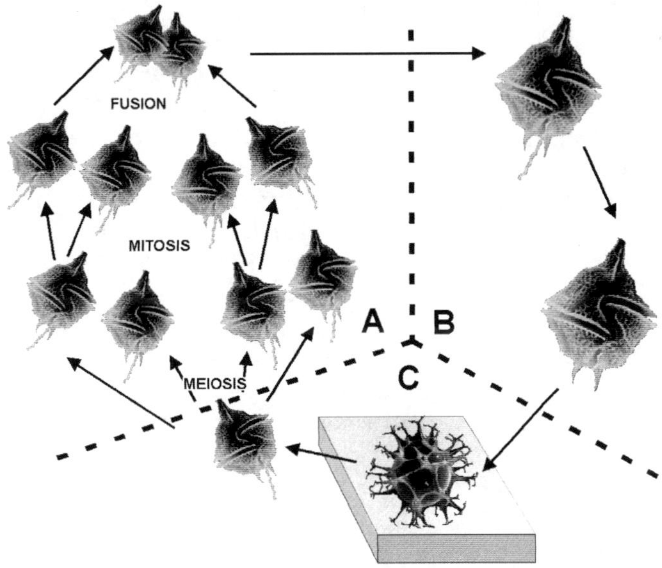

Figure 1. Schematic dinoflagellate life cycle. A) Haploid phase where individual cells divide through mitosis. This is the vegetative stage, which leads to the formation of bloom events. Following an internal and/or external stimuli, gametes are formed and will fuse to produce a diploid zygote; B) Planozygote (pelagic motile zygote) phase where the diploid cell will eventually lose motility, and in which the cyst will develop and eventually be released in the water column, then down to the sediment surface; C) Hypnozygote stage where the cyst will undergo a mandatory dormancy period controlled by an endogenous clock before excystment takes place. Modified after Evitt [31].

The distribution of dinoflagellates is not only dependent on the physico-chemical variables (currents, temperature, salinity, irradiance, nutrients) of the water column, but also on their feeding strategies, which may include mixotrophy and phagotrophy (ingestion of large food particles or preys) as well as the distribution of their prey. Over the last decade, several research programs aimed at studying the changing Arctic environment were conducted in the northern Baffin Bay (International North Water Polynya Study), the Beaufort Sea (Canadian Arctic Shelf Exchange Study) and the Canadian Arctic Archipelago (ArcticNet). The ecology of plankton and ice diatoms and other protists have been extensively studied over long-term periods as part of these Arctic research programs, but very little has been done on the ecology of dinoflagellate bloom events. The main reasons are that dinoflagellates are usually less abundant in the water column and they are adapted to growth under low turbulence and low nutrients concentrations [56]. Dinoflagellate blooms often occur after the one of diatoms and usually are restricted in time and space, therefore easily missed. For example, in highly productive Arctic basins, such as the North Water Polynya in northern Baffin Bay, diatoms usually dominate the phytoplankton and may account up to 70% of the total phytoplankton carbon biomass, while dinoflagellates and ciliates comprise approximately 24% of the total phytoplankton carbon biomass [24, 32]. However, dinoflagellates may account for the majority of the phytoplankton carbon biomass during certain periods in relation to the depth of the surface mixed layer.

2.1. Cyst-theca relationships of Arctic dinoflagellates

Dinoflagellates have been studied by biologists since the mid-late eighteenth century, while dinoflagellate cysts were studied by marine palynologists since the early nineteenth century [31], which lead to a double nomenclature system for the motile and cyst stages. The first cyst incubation experiments in the 1960s [33] paved the way to a number of studies aiming at establishing the cyst-theca relationship of the cyst stage [34, 35, 36, 37, 38, 39]. Several dinoflagellate species closely resemble each other and, in several cases, only detailed examination in scanning electron microscopy resulted in accurate identification. The best example is the *Gonyaulax spinifera* (Claparède *et* Lachmann 1859) Diesing 1866 complex, in which ~13 cyst species were supposedly derived from one single motile species [39]. Therefore, species identification can sometimes be erroneous when observing plankton samples because the minute details may have been missed.

Six studies provide checklists of motile dinoflagellate species present in the Canadian High Arctic [18, 23, 24, 27, 28, 29] and they accounted for most of the phytoplankton surveys in the area since the 1970s. A total of ~192 taxa belonging to 42 genera have been identified, among which 10 species are known to produce resistant resting cysts as part of their life cycle (*Brigantedinium cariacoense* = *Protoperidinium avellana* (Meunier 1919) Balech 1974; *B. simplex* = *Protoperidinium conicoides* (Paulsen 1905) Balech 1973; *Pentapharsodinium dalei* Indelicato *et* Loeblich 1986 = theca-based name; *Polykrikos kofoidii* Chatton 1914 = theca-based name; *Polykrikos schwartzii* Bütschli 1873 = theca-based name; *Protoperidinium americanum* (Gran *et* Braarud 1935) Balech 1974 = theca-based name; *Protoperidinium nudum* = theca-based name; *Spiniferites elongatus* = *Gonyaulax elongata* (Reid 1974) Ellegaard, Daugbjerg, Rochon, J. Lewis *et* Harding 2003; *Trinovantedinium applanatum* = *Protoperidinium pentagonum* (Gran 1902) Balech 1974). In some cases, the cyst has been observed while forming inside the theca within plankton samples, while in other cases, the relationship has been confirmed through laboratory cyst incubation experiments. Several studies have recorded a total of ~31 dinocyst taxa from the Beaufort Sea to Baffin Bay [14, 40, 41, 42, 43, 44, 45]. In this case, the cyst-theca relationships were confirmed for 10 species only [1], which means that the plankton motile stage of over 67% of the recorded cysts are still unknown. Furthermore, the motile equivalent of only 2 of these 10 species have been observed in Arctic plankton samples so far, which leaves us with ~94% of Arctic dinocyst species for which the motile counterparts have not been observed in the area where they were found.

3. Implication for paleoceanographical reconstructions

The fact that only 6% of Arctic dinocyst species are linked to their motile stage has important consequences for both biologists and paleoceanographers. From a biological point of view, it implies that plankton surveys are still incomplete and that potentially important dinoflagellate species are not taken into account in Arctic plankton ecological and food web studies. From a paleoceanographical point of view, the concerns are numerous. Dinoflagellate cyst assemblages are routinely used as a proxy indicator of conditions in the upper part of the water column [46] in a number of studies on the evolution of marine environments in the Holocene (last 10,000 years) period [3, 4, 44, 47, 48]. These studies relied on the relationships between the modern dinoflagellate cyst assemblages in surface sediments samples and the upper water column variables measured above each sampling site. The cyst-theca relationships will provide valuable information on the ecological requirements of the motile and cyst stages, which will in turn improve the paleoceanographical interpretations. This is especially relevant since some of the key species used for paleoceanographical reconstructions, which represent a significant proportion of the dinocyst assemblages (e.g., *Islandinium minutum, Spinferites frigidus, Echinidinium karaense, Polykrikos* sp.), are still not linked to a motile and vegetative dinoflagellate stage. Such studies are presently being performed on some of these ecologically important dinocyst species [49] with the aim to document the ecological requirements of the motile stage. In one case, the motile form was actually a new dinoflagellate species that is in presently being described. Plankton surveys are also being done in order to document the occurrence of these species that have not been documented yet.

There are also issues regarding the transport of dinocysts prior to, or after, their deposition in the surface sediments. Indeed, dinoflagellate cysts are in the size range of silt particles (4 to 63 μm) and behave as such [50]. Between the time they are formed in the water column and settled down in the surface sediments, dinocysts are subject to horizontal transport by surface and/or subsurface currents. This implies that several of the cysts retrieved in the surface sediments may have been formed elsewhere, and are therefore not representative of the upper water column conditions immediately above the collection site. A study showed that most dinocyst assemblages found in the deeper parts of oceanic basins were in fact the result of long-distance transport [51]. However, other studies have indicated that the dinocyst assemblages found in relatively shallow coastal areas (~ 500 m) were representative of the local/regional dinoflagellate populations in the overlying water column [52, 53]. It is therefore assumed that this may also be the case for the Canadian High Arctic where most sampling sites are located in relatively shallow areas.

Another point to consider is how accurately surface sediment samples represent the modern dinocyst assemblages and the upper water column conditions. The sediment accumulation rates vary greatly from one region of the Canadian High Arctic to another. Most shelf areas and channels of the Canadian Arctic Archipelago are characterised by low sedimentation rates due to the lack of sediment sources. High sediment accumulation areas are restricted to localized basins. For example, in the Mackenzie Trough [3, 54] the sedimentation rate can reach 0.12 cm per year, while in the nearby Amundsen Gulf, it is < 0.001 cm per year [55]. In the eastern Canadian High Arctic, the sedimentation rate measured 0.17 and 0.44 cm per year just outside of Jones Sound and in Hudson Strait respectively [54]. Therefore, the upper 1 cm of sediment that is routinely collected and analysed for "modern" dinocyst assemblages may represent between 10^0 and 10^3 years. This implies that some of the dinocysts recovered from the surface sediments may in fact be indicative of motile dinoflagellate populations that were no longer present in the area. However, this is unlikely for the majority of the dinocyst species found, since repeated surface sediment sampling campaigns over the last 10 years have yielded relatively important concentrations of viable cysts with cell content.

In spite of these limitations, dinocyst assemblages constitute an excellent tracer of past upper water column conditions. However, a better knowledge of their biological affinities and their relation to the phytoplankton assemblages and sea surface conditions can only enhance the paleoceanographical interpretations made from fossil cyst assemblages. Part of this knowledge will be acquired through incubation experiments on Arctic dinocyst species.

References

[1] Head MJ 1996 Chapter 30. In Palynology: Principles and Applications **3** (AASP Foundation) 1197-1248

[2] Dale B 1983 Survival Strategy of the Algae (Cambridge University Press) 69-136

[3] Richerol T, Rochon A, Blasco S, Scott DB, Schell TM and Bennett RJ 2008 *Mar. Micropal.* **68** 6-20

[4] Ledu D, Rochon A, de Vernal A and St-Onge G *Can. J. Earth Sci.* In Press

[5] de Vernal *et al.* 2006 *Quat. Sci. Rev.* **25** 2820-2834

[6] Harland R, Nordberg K and Filipsson HL 2006 *Sci. Total Env.* **355** 204-231

[7] Vásquez-Bedoya LF, Radi T, Ruiz-Fernández AC, de Vernal A, Machain-Castillo ML, Kielt JF and Hillaire-Marcel, C 2008 *Mar. Micropal.* **68** 49-65

[8] Thibodeau B, de Venral A and Mucci A 2006 *Mar. Geol.* **231** 37-50

[9] Radi T and de vernal A 2008 *Mar. Micropal.* **68** 84-114

[10] Radi T, Pospelova V and de Vernal A 2007 *Mar. Micropal.* **62** 269-297

[11] de Vernal *et al.* 2005. *Quat. Sci. Rev.* **24** 897-924

[12] Marret F and Zonneveld K 2003 *Rev. Palaeobot. Palynol.* **125** 1-200

[13] Matthiessen J 1995 *Mar. Micropal.* **24** 307-334

[14] Rochon A, de Vernal A, Turon JL, Matthiessen J and Head MJ 1999 *AASP Cont. Ser.* **35** 1-146

[15] Kunz Pirrung M 2001 *J. Quat. Sci.* **16** 637-650

[16] Radi T and de Vernal A 2004 *Rev. Palaeobot. Palynol.* **128** 169-193

[17] Vorlova TY, Morozova TY, Gribble KE, Kulis, DM and Anderson DM 2004 *Bot. Mar.* **47** 184-201

[18] Hsiao SIC 1983 *Nova Hedwigia* **37** 225-313

[19] Okolodkov YB and Dodge JD 1996 *J. Exp. Mar. Biol. Ecol.* **202** 19-27

[20] Okolodkov YB 1998 *Sarsia* **83** 267-292

[21] Okolodkov YB 1999 *Grana* **38** 162-169

[22] Okolodkov YB, Hapte R and Semovski SV 2000 *Sarsia* **85** 345-352

[23] Booth BC and Horner RA 1997 *Deep-Sea Res. II* **44** 1607-1622

[24] Lovejoy C, Legendre L, Martineau MJ, Bâcle J and von Quillfeldt CH 2002 *Deep-Sea Res. II* **49** 5027-5047

[25] Vershinin A, Moruchkov A, Morton SL, Leighfield TA, Quilliam MA and Ramsdell JS 2006 *Harmful Algae* **5** 558-564

[26] Gómez F 2005 *Acta Bot. Croat.* **64** 129-212

[27] Riedel A, Michel C, Poulin M and Lessard S 2003 Canadian Data Report of Hydrography and Ocean Sciences **159** 54pp

[28] Różańska M, Poulin M and Gosselin M 2008 *J. Mar. Sys.* **74** 887-901

[29] Różańska M, Gosselin M, Poulin M, Wiktor JM and Michel C *Mar. Ecol. Prog. Ser.* Submitted

[30] Kokinos JD, Eglinton TI, Goni MA, Boon JJ, Martoglio PA and Anderson DA 1998 *Org. Geochem.* **28** 265-288

[31] Evitt WR 1985 *Sporopollenin Dinoflagellate Cysts* (AASP Fundation) 333pp

[32] Vidussi F, Roy S, Lovejoy C, Gammelgaard M, Thomsen HA, Booth B, Tremblay J-E and Mostajir B 2003 *Can. J. Fish. Aquat. Sci.* **61** 2038-52

[33] Wall D and Dale B 1968 *Micropal.* **14** 265-304

[34] Matsuoka K 1984 *Bulletin of the Faculty of Liberal Arts, Nagasaki University, Natural Science* **25** 37-47

[35] Matsuoka K 1985 *Transactions and Proceedings of the Paleontological Society of Japan, New Series* **140** 240-262

[36] Lewis J 1990 *Brit. Phycol. J.* **25** 339-351

[37] Lewis J 1991 *Bot. Mar.* **34** 91-106

[38] Lewis J, Dodge JD and Tett P 1984 *J. Micropal.* **3** 25-34

[39] Lewis J, Rochon A and Harding IC 1999 *Grana* **38** 113-124

[40] Harland R, Reid PC, Dobell P and Norris G 1980 *Grana* 211-225
[41] Mudie PJ and Rochon A 2001 *J. Quat. Sci.* **7** 603-620
[42] Solomon S, Mudie PJ, Cranston R, Hamilton T, Thibeaudeau SA and Collins ES 2000 *Int. J. Earth Sci.* **89** 503-521
[43] Richerol T, Rochon A, Blasco S, Scott DB, Schell TM and Bennett R 2008. *J. Mar. Sys.* **74**, 825-839
[44] Hamel D, de Vernal A. Gosselin M and Hillaire-Marcel C 2002 *Deep Sea Res. II* **49** 5277-95
[45] Rochon A and de Vernal A 1994 *Can. J. Earth Sci.* **31** 115-127
[46] de vernal A et al 2005 *Quat. Sci. Rev.* **24** 897-924
[47] Levac E, de Vernal A and Blake W 2001 *J. Quat. Sci.* **16** 353-363
[48] McKay J, de Vernal A, Hillaire-Marcel C, Not C, Polyak L and Darby D *Can. J. Earth Sci.* In Press
[49] Rochon A and Potvin E. *Dino 8, Conference and Abstract* p52
[50] Dale B 1976 *Rev. Palaeobot. Palynol.* **22** 39-60
[51] Dale B 1992 *Chapter 1, Ocean Bioscenosis Series* **5** 1-21
[52] Reid PC 1975 *New Phytol.* **75** 589-603
[53] Anderson DM, Stock CA, Keafer BA, Nelson AB, McGillicuddy DJ, Keller M, Thompson B, Matrai PA and Martin J 2005 *Deep-Sea Res. II* **52** 2522-42
[54] Rochon A, Scott DB, Schell TM, Blasco S, Bennett R and Mudie PJ 2006 *AGU Spplement and Abstract* U43B-0867
[55] Schell TM, Moss TJ, Scott DB and Rochon A 2008 *J. Geophys. Res.* **113** C03S02 doi: 1029/2007JC004202
[56] Margalef R 1978 Life-forms of phytoplankton as survival alternatives in an unstable environment. *Oceanol. Acta* **1** 493-509

Calcareous nannoplankton and benthic foraminiferal assemblages from the Nazaré Canyon (Portuguese continental margin): preliminary results

C Guerreiro[1], F Rosa [2], A Oliveira[1], M Cachão[3], F Fatela[3], and A Rodrigues[1]

[1] Division of Marine Geology, Portuguese Hydrographic Institute (IH), Rua das Trinas 49, 1249-093 Lisboa, Portugal

[2] CIACOMAR, Algarve University, Av. 16 de Julho s/n 8700-311 Olhão, Portugal

[3] Geology Center and Geology Department, FCUL, Bloco C6, 3° Piso, sala 6.3.57 Campo Grande 1749-016 Lisboa, Portugal

E-mail: catarina.guerreiro@hidrografico.pt

Abstract. Submarine canyons are assumed to play an important role in oceanic/neritic circulation, marine productivity and sedimentary processes, acting as preferential conduits between the littoral and deep oceanic domain. Here we present first results of a comparative micropalaeontological study on calcareous nannoplankton and benthic foraminifera from surface sediments from the surroundings of the upper Nazaré Canyon (Portuguese continental margin) and from the shelf north of the canyon. Regardless of the difficulty to distinguish taphonomical from (palaeo)ecological effects in such a complex and still poorly known marine system, the first results suggest that the canyon's hydro-sedimentary dynamic regime act as a prolongation of the shore/inner shelf hydrodynamic conditions towards west, preventing deposition and/or preservation of the smaller and fragile species of calcareous nannoplankton (*e.g. E. huxleyi* and *G. ericsonii*) and enhancing the record of the larger and more opportunistic ones (*e.g. G. oceanica*); and disturbing benthic foraminiferal productivity and/or diversity, or their preservation in the fossil record. Both calcareous nannoplankton and benthic foraminifera are more abundant off the canyon's domain, suggesting that its highly energetic thalweg conditions are probably filtering the fossil record in the sediment. Still, preliminary results suggest that the occurrence of persistent physical phenomena related with the canyon's morphology and proximity to the coast (*e.g.* solitary internal waves) may be locally promoting favourable conditions for calcareous nannoplankton, as shown by high values of nannoliths, chlorophyll *a* and 19' hexanoyloxyfucoxantine (unpublished data) north of the canyon's head. It is our goal to test this hypothesis in the near future by (a) studying multicore and surficial sediments from more recent surveys, and (b) calibrating the sediment results with water column data presently in process at the Institute of Oceanography (IO).

1. Introduction

Active submarine canyons are major sedimentary conduits between the coastal zone and the deep oceanic domain, affecting both oceanic productivity and circulation patterns, and also influencing the sedimentary cover of the adjacent shelf and slope. Their efficacy is strongly dependent on the local and regional energetic conditions (*e.g.* upwelling, internal waves, and turbidity currents), the distance between the canyon head and the coastline, the canyon morphology and the geological features of the adjacent shelf.

The Nazaré Canyon, a gouf-type canyon, is one of Europe's largest submarine valleys, located on the central part of the Portuguese margin and orientated roughly perpendicular to the coast in an E – W direction. The canyon lies between 39°20'N and 40°00'N with a length larger than 250km cutting across the entire shelf almost to the beach and is not connected to a major river system [1]. Its head indents the coastline at a point where the coast shifts from an N – S to a NE – SW orientation, probably capturing most of the long-shore sediment transport [2].

The upper part of the Nazaré Canyon is a sharp V-shaped valley deeply incised in the shelf and steep upper slope, descending from the shore to 2700m depth over a length of 70km measured along the canyon axis. *Vitória valley* is one short side-valley that branches off from the upper canyon in NE direction.

Sediments recovered from the Nazaré Canyon consist mostly of terrigenous mud, in some parts of the thalweg covering sand of presumed turbiditic origin [3] [4] [5]. The shelf south of the canyon is covered largely with rocky outcrops and coarse-grained sediments whereas finer sand and mud prevail on the shelf on the northern side [6] [5]. According to [5] material derived from three different sources is captured by the Nazaré Canyon: (1) material resuspended from the middle shelf by internal wave activity and transported to the canyon in the BNL during upwelling conditions; (2) material eroded from northern/southern beaches and sea cliffs transported by littoral drift to the canyon head; and (3) fine sediment supplied by the southern rivers and transported to the canyon during downwelling conditions (winter).

[3] has demonstrate that fine-grained suspended sediment from the adjacent shelf is captured in the upper canyon and actively transported through the canyon to the abyssal plain under the influence of tidal currents and intermittent sediment gravity flows.

Investigation performed by [7] provided evidence that both internal waves as well as episodic sediment gravity flows [3] [4] are active in transporting sediment though the canyon. Additionally, very high rates of sediment accumulation have been reported from the upper and middle part of the canyon, one or two orders of magnitude more than at comparable depths on the open slope [8] [4]. These high rates are probably due the fact that the Nazaré Canyon cuts the full width of the shelf, and therefore intercepts a large portion of the sediment transported laterally along the shelf [4].

In order to understand the role played by the submarine canyon's morphology, location and hydrodynamic complexity in the spatial distribution of benthic and planktonic marine microorganisms, a comparative micropalaeontological study was performed on the fossil record of calcareous nannoplankton and benthic foraminifera from surficial sediments. Samples were collected in the surroundings of the upper Nazaré Canyon and from the shelf north of the canyon unaffected by canyon processes.

2. Material and Methods

Absolute abundances of calcareous nannoplankton were determined in 69 surface sediment samples from the Nazaré Canyon region (EUROSTRATAFORM survey performed by Portuguese Hydrographical Institute, in May 2004) (Figure 1). Slides were prepared according to the random settling procedure [9] and observed under optical polarizing microscope (Olympus BX-40), at 1250× magnification. A minimum of 300 individual nannoliths were counted and identified in each slide to characterize calcareous nannoplankton assemblages and determine their absolute abundances, expressed as nannoliths per gram (nanno/g). A statistical multivariate analysis was performed over the studied samples (69 cases) with calcareous nannoplankton species abundances (nanno/g) as variables (11 taxa), using the method of Principal Factors (Communalities = multiple R^2) of *Statistica* 7 analytical software. Results from the original data matrix were optimized through a two axes projection using a Varimax Raw rotation. Four main factors were extracted and their score values were spatially represented in isoline maps using ordinary krigging (contour map, *Surfer 8*).

Relative abundances of benthic foraminiferal species were determined in 29 samples (SEPLAT program - cruises 20 (November 1995), 21 (October 1996) and 23 (October 1998), IH) selected from the submarine canyon (between 74 m and 458 m depth) and from the shelf north of the canyon (between 20 m and 170 m depth) (Figure 1) where [10] registered maximum values of foraminifera shells in sand-rich sediments. Samples were washed over a sieve and the fraction >63μm was observed under a binocular microscope. Total assemblages were considered because it was not possible to distinguish living from dead foraminifera. A minimum of 300 individuals were counted and identified in each sample. Identification of foraminifera species followed revised classification by [11] and diversity of species was obtained with the Shannon-Wiener Index (H). A multivariate statistical analysis (*Statistica 7*) was performed on combined benthic foraminiferal species distribution and granulometric parameters, to investigate correlations between the microfauna's and abiotic factors.

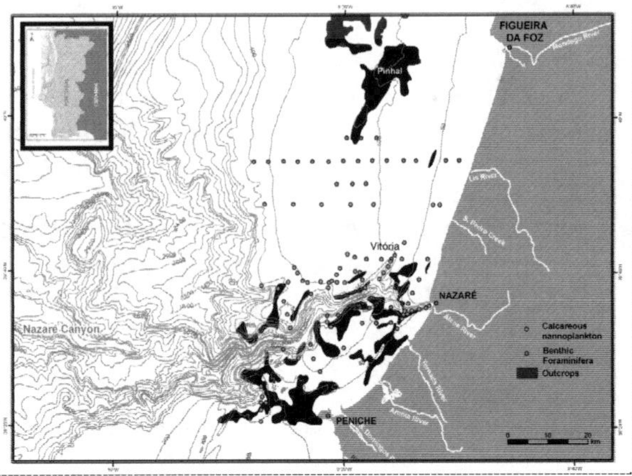

Figure 1. Bathymetric map of the Nazaré Submarine Canyon and location of surface sampling stations: EUROSTRATAFORM (blue) and SEPLAT-IH (red).

TOC (Total Organic Carbon) was determined in 67 samples (EUROSTRATAFORM) by infrared absorption spectrometry using STROHLEIN Cmat 5500 equipment, and grain-size (Silt–Clay fraction percentage) was determined using a Malvern master sizer 2000G).

3. Results

3.1. Calcareous nannoplankton

Diversity in the Nazaré Canyon is identical to that found in other sectors of the Portuguese shelf [12] with abundance values typical of the neritic domain, although occasionally close to low values typically found in the littoral domain. The complex distribution pattern of the abundance values is also reflecting that proximity. High amounts of temperate (*Gephyrocapsa muellerae*) and subtropical (*Gephyrocapsa oceanica* and *Calcidiscus leptoporus*) species is concordant with the location of the Portuguese margin in the transition from the Temperate to the Subtropical Biogeographic Zone [13]. Small species (*Emiliania huxleyi* and *Gephyrocapsa eriscsonii*) dominate the assemblages in most of the studied samples.

Multivariate analysis reveals that four main species assemblages characterize the studied sediment samples (Figure 2); *Factor 1* is represented by *E. huxleyi* and *G. ericsonii* (nannoliths <3μm), preferentially distributed along the canyon's northern margin, some distance away from both the littoral zone and the more coastward part of the upper canyon thalweg, (eastwards of the outer shelf) (Figure 2a); *Factor 2* is reflecting the positive local anomalies of some rare species, namely *Syracosphaera* spp., *Syracosphaera pulchra, Braarudosphaera bigelowi* and *Umbilicosphaera sibogae-foliosa*, unexpectedly recorded close to the shore on the inner shelf north of the canyon head, and in three areas systematically characterized by high diversity and abundance of nannoliths: (i) in the canyon head; (ii) at 39°05'N 009°05'W on the northern canyon edge near a prominent bend in the canyon axis, and (iii) in the canyon's southern tributary (Figure 2b); *Factor 3* is represented by *Helicosphaera carteri, Coccolithus pelagicus, Calcidiscus leptoporus* and *Gephyrocapsa muellerae,* which have a geographical distribution largely similar to *Factor 1, i.e.* preferentially concentrated on the canyon's northern margin but also in the hydrodynamically active littoral zone and canyon's proximal sector (Figure 2c); Finally, *Factor 4* is mainly represented by *G. oceanica,* the distribution of which is preferentially bordering the canyon's slopes, where other species are rare or completely absent (*e.g. E. huxleyi* and *G. ericsonii* are absent in certain areas of the canyon's southern margin) (Figure 2d).

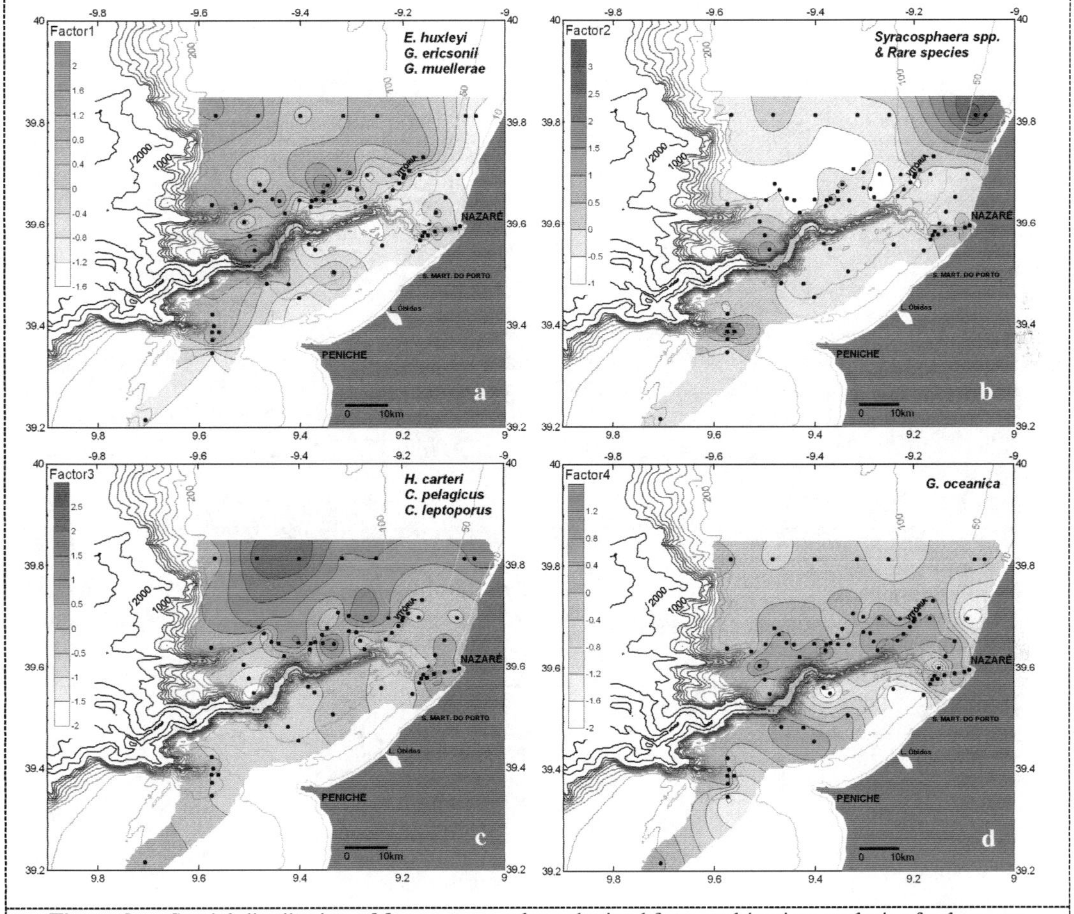

Figure 2. Spatial distribution of factor score values obtained from multivariate analysis of calcareous nannoplankton assemblage data: a = Factor 1; b = Factor 2; c = Factor 3 and d = Factor 4

Large calcareous nannoplankton species (dimension $\geq 3\mu m$) from most of the studied samples have abundance values between 4.6×10^6 e 4.8×10^6 nanno/g that remain quite stable independent of the silt-clay fraction and TOC. Minor species (measuring $<3\mu m$; *E. huxleyi* and *G. ericsonii*) also reveal this independence with silt-clay fraction and TOC values (Figure 3). The nannoliths' positive anomaly observed on the inner shelf (represented by *Factor 2 scores and* rose rectangle in Figure 3) correspond, in general, to fine sands with high calcareous nannoplankton abundance and diversity, and very low values of silt-clay fraction (<< 10%) and TOC (<0.5).

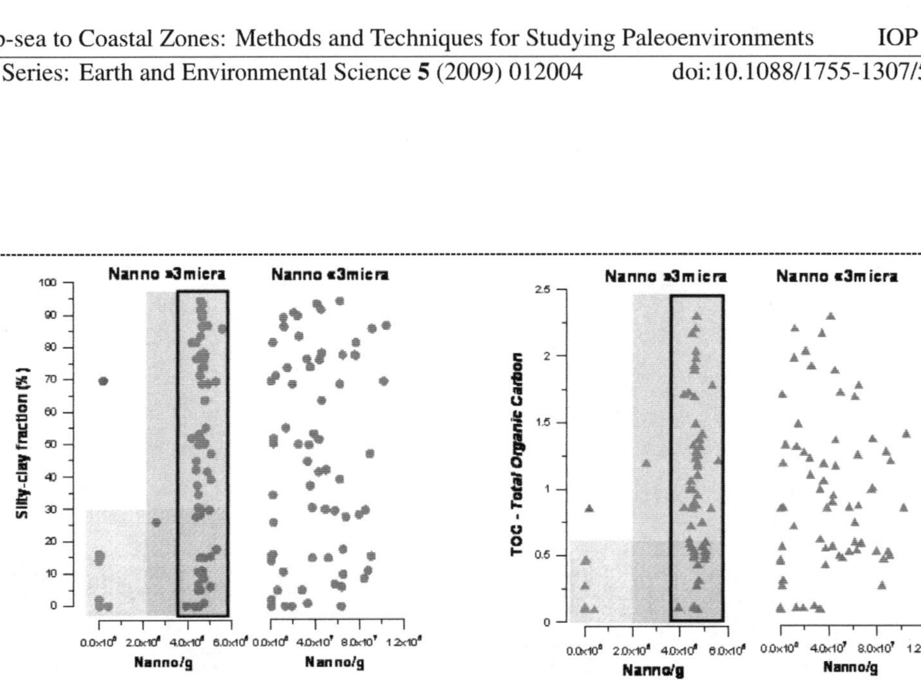

Figure 3. Silt-Clay fraction (%) and TOC *versus* nannolith's abundance. Black line rectangle= high and constant values of nannolits, independently of Silt-Clay fraction and TOC content; green rectangle = high values of nannoliths, Silt-Clay fraction and TOC; rose rectangle= high values of nannoliths and low values of both Silt-Clay fraction and TOC; orange rectangle= low values of the three studied parameters.

3.2. Benthic foraminifera

Benthic foraminiferal abundances in the sediment are higher on the shelf than near the submarine canyon. In both domains foraminiferal abundance is usually higher in more distal environments further offshore, although in the canyon there is also a clear trend for higher abundances on and around the canyon terraces than in the axis. Species common to both domains usually occur at similar depths and related to the same type of sediment.

Diversity (H) of species is slightly higher on the shelf (H ca. 3.5 to 4) than in the Nazaré Canyon domain (H ca. 3 to 3.5), where it remains constant at almost every depth (Figure 4).

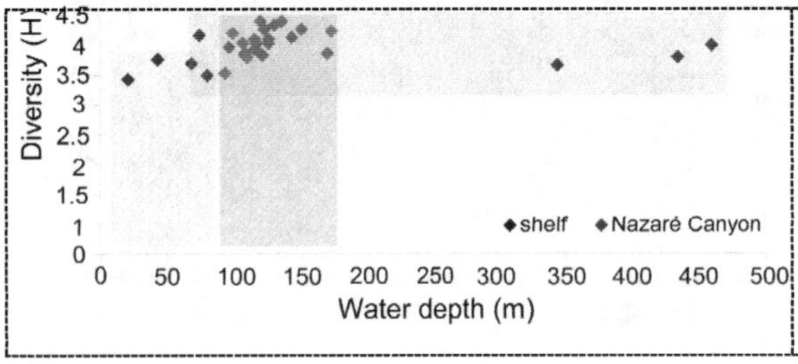

Figure 4. Diversity index H (Shannon-Wiener) of benthic foraminifera in the northern shelf and in the Nazaré Canyon.

On the shelf domain, H varies in good correlation with bathymetry, increasing below 100-120 m water depth, in the transition from inner/middle shelf towards the outer shelf (Figure 4).

The distribution (presence/absence and relative abundance) of foraminiferal species also reveals some significant differences between the continental shelf and the Nazaré Canyon (Figure 5), and also show a strong correlation with sediment type. On the continental shelf until about 80 m depth, where coarser sediments (sand and gravel) prevail, dominant species are *Bulimina elongata* (BELONG), *Cibicides lobatulus* (CLOBAT), *Gavelinopsis praegeri* (GPRAEG), *Discorbinella berthelotti* (DBERTH), *Planorbulina mediterranensis* (PMEDITER), *Textularia conica* (TCON) and *Cassidulina crassa* (CCRASS). Between 80 m and 120 m depth, where the highest abundances of silt-clay fraction in the continental shelf were observed, dominant species are *Bolivina dilatata* (BDILAT), *Brizalina spathulata* (BRIZSPA), *Bulimina marginata* (BMARG) and *Cassidulina laevigata* (CLAEV), and an increasing abundance is found for *Nonionella turgida* (NTURG), *Epistominella exigua* (EEXIG) and *Uvigerina peregrine* (UPEREG). Below 120 m depth until the shelf edge (180-200 m depth), the same species maintain their dominance but *Bulimina striata* and *Globocassidulina subglobosa* (GSUBGLOB) gain importance as accessory species.

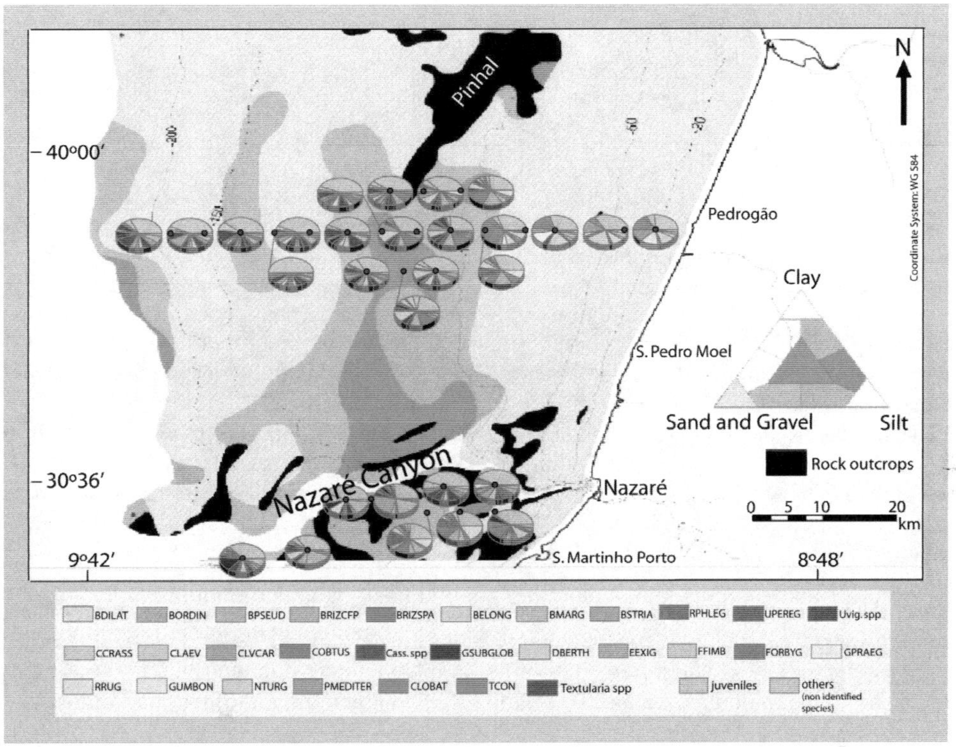

Figure 5. Spatial distribution of benthic foraminiferal species (% in total assemblages). Sedimentological data from Pombo (2004)

In the Nazaré Canyon, species distribution reflects the existence of a microfaunistic boundary around 150 m depth. Shallower areas of the canyon, above 150 m depth, are dominated by *Brizalina spathulata*, *Bolivina dilatata*, *Uvigerina peregrina* and *Rectuvigerina phlegeri*, and *Textularia conica* and *Cibicides lobatulus* are also present in fair numbers. The latter, together with *Cassidulina crassa*, are dominant in the shallowest area of the canyon. In the deeper areas of the canyon, below 150 m depth, the most abundant species is *Bolivina pseudoplicata* (BPSEUD). *Bulimina striata* (BSTRIA), *Gyroidina umbonata* (GUMBON) and *Globocassidulina subglobosa* are exclusively found in this area. Factor analysis (Figure 6) applied to relative abundance of benthic foraminiferal species and texture of the sediment (silt-clay percentage in total sediment) shows

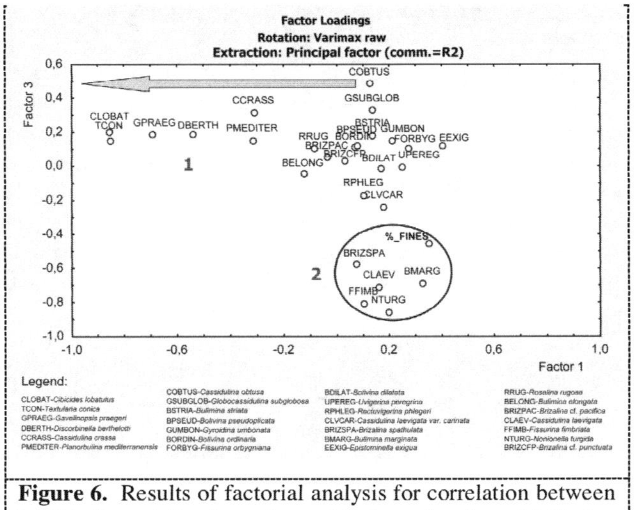

Figure 6. Results of factorial analysis for correlation between benthic foraminiferal species and texture of sediment (silt-clay percentage in total sediment).

that the first factor explains ca. 24% of total variance, the second ca. 13% and the third ca. 9.5%. Together they explain 65% of the total variance observed in the microfaunistic and grain-size characteristics of the sediments.

4. Discussion

4.1. Taphonomical versus (Palaeo)ecological record of calcareous nannoplankton

"Gouf" type submarine canyons are known to act as preferential conduits for oceanic water masses towards more proximal near-coast domains, and thus an increase in calcareous nannoplankton abundance and diversity along the canyon's slopes was expected, as compared to the adjacent shelf. However, results reveal the exact opposite; nannoliths are more abundant and diversified on the shelf north of the canyon, away from the area influenced by canyon processes, suggesting that the Nazaré Canyon's complex and highly active hydrodynamic conditions are preventing their deposition and preservation inside the canyon.

The record of calcareous nannoplankton in the sea bottom sediment (nanno/g) usually represents the resultant from the occurrence (and intensity) of three main environmental factors: (a) phytoplankton productivity in the water column; (b) preferential settling and sorting, merely physically controlled and/or enhanced by biogenic forcing (*e.g.* zooplankton pellets); (c) diagenetic dissolution in the water column or inside the sediment; and (d) dilution with other sedimentary material.

The dissolution degree (c) is usually evaluated by the relative amount of the more vulnerable small nannolith species (*e.g. G. ericsonii* and *E. huxleyi*) generally used as an indicator of the quality of the nannoliths assemblages. Since these species are dominant (relative abundance >70%, ~10^8nanno/g) in 82% of the studied samples, we assume that the studied assemblages were not subjected to significant dissolution.

Enhancement of the nannolith record by preferential settling (b) can be evaluated by its correlation with the relative percentage of silt-clay fraction in the sediment. Thus, samples with high nannolith abundance but reduced silt-clay fraction must reflect other factors such as consistent favorable conditions for the productivity of calcareous nannoplankton or sediment starvation. TOC would be a good indicator of the marine productivity, but according to [5] it has a predominantly continental signature in the studied area, and a distribution closely similar to that of the silt-clay fraction, *i.e.*, apparently dependent on the hydrodynamical setting.

Distribution patterns observed in calcareous nannoplankton assemblages of the Nazaré Canyon area are complex, reflecting the sum of environmental variables (sedimentological and physical) interacting in this highly complex and still poorly understood (in what concerns the phytoplankton ecology) area. Results are discussed and interpreted, integrating the present-day knowledge of this marine system in what concerns its sedimentary [4] [3] [5] [1] and physical processes [7] [14] (Vitorino *et al.*, in prep.).

The clear dominance of *E. huxleyi* and *G. ericsonii* in most of the studied samples is reflected by *Factor 1*, together with *G. muellerae*, which represents the open ocean signal in the studied assemblages [12] [15]. As expected, *Factor 1* scores low close to shore, probably related to both higher hydrodynamic activity preventing settling of these small species and unfavorable conditions for development of oceanic calcareous nannoplankton. The same effect seems to occur close to the canyon suggesting that the canyon has a comparable effect on nannolith distribution as the coastline border.

Species of both *Factor 1* and *3* represent the typical calcareous nannoplankton assemblages and abundances to be found in the "normal" (*i.e.* not affected by canyon processes) Portuguese continental shelf [12]. Some species scoring on *Factor 3* (*H. carteri, C. pelagicus* and *C. leptoporus)* revealed to be better adapted to more coastal areas probably due their larger cell size or opportunistic behavior, which is in agreement with earlier palaeoenvironmental studies performed in the Portuguese continental shelf [12] [15].

To the contrary, *Factor 2* is registering particular anomalous nannolith occurrences; some of these may be reflecting the existence of local areas of low hydrodynamic activity apparently related with the canyon's morphology, that may act as depocenters of silt-clay particles, both terrigenous and biogenic (*e.g.* nannoliths). However, the nannolith positive anomaly found on the inner shelf north of the canyon head, may have ecological significance; Coccolithophores generally prefer more oceanic and oligotrophic conditions [16], and inner shelf sediments are mostly sandy (silt-clay percentage <10%, Figure 3), indicative of unfavorable hydrodynamic conditions for settling of nannoliths. Thus, the high abundance of nannoliths must be explained by other physical mechanisms, promoted or enhanced by the canyon's morphology and proximity to the littoral, *e.g.* intensification of upwelling [17]; generation of internal waves and tides [7].

Recent biological sampling (HERMES cruise performed by Portuguese Hydrographic Institute, June and July of 2006) presently in study in the Institute of Oceanography (IO) in the Nazaré Canyon during upwelling conditions recorded enhanced values (ugL^{-1}) of both chlorophyll ***a*** (phytoplankton biomass index) and 19' hexanoyloxyfucoxantine (pigment indicator of coccolithophores) in this same sector [14]. This suggests that the high abundance and diversity of nannoliths found in this area may indeed be registering the occurrence of a persistent productivity event in this part of the shelf. The internal solitary waves generated in the canyon's northern shelf break and refracted towards NE [7] probably could explain this record, by forcing the injection of nutrients from the lower toward the upper levels of the water column (Vitorino, 2008, *personal comun.*). Although this mechanism occurs mainly during spring/summer, it seems persistent enough (semidiurnal) to explain the observed anomaly. High nannoliths abundance and low TOC content in this area [5] could be explained by a rapid transport of the nannoliths towards the bottom, probably within zooplankton faecal pellets (*e.g.* Copepods) followed by post depositional fast organic matter oxidization.

Despite the subtropical character of *G. oceanica* (*Factor 4*), results are in agreement with earlier studies showing that this species is usually well adapted to more coastal regions, possibly because of its larger dimensions and relative cosmopolitism [18] [12]. Its relatively higher abundance along the canyon's northern border is another indication that the canyon's hydrodynamic regime is comparable to that of any other coastal area, by enhancing *G. oceanica* in the sedimentary record.

The overall lack of correlation between the larger nannoliths' abundances (nanno/g $\geq 3\mu m$) and both TOC and silt-clay fraction (%) contained in the majority of the samples (Figure 3), and the high values of the smaller species (*e.g. E. huxleyi* and *G. ericsonii*, with max. 1.8×10^8 nanno/g) dominating most assemblages argue in favour that the calcareous nannoplankton record is not strictly dependant on local hydrodynamic setting but also contains a (palaeo)ecological signal. Prevailing southward upwelling currents may also have an important role in the distribution pattern of calcareous nannoplankton.

4.2. *Benthic foraminifera in relation with the sedimentary cover*

The presence and/or abundance of certain benthic foraminiferal species reveals a strong correlation with the textural characteristics of the sediment cover, probably because the same environmental parameters that are controlling depositional patterns are also of importance for the ecology and preservation of benthic foraminiferal species.

Our data show that for benthic foraminifera which occur in both domains, continental shelf and Nazaré Canyon, there are two groups of species with different behaviour in what concerns colonization of different type of sediment: (1) *Cibicides lobatulus, Textularia conica, Gavelinopsis praegeri, Discorbinella berthelotti, Planorbulina mediterranensis* and *Cassidulina crassa* seem to be found preferably depositional environments characterized by coarser sediments, which tend to be more abundant in shallower areas of both domains; and (2) *Nonionella turgida, Cassidulina laevigata, Bulimina marginata* and *Brizalina spathulata* show a strong affinity with fine sediments, which are generally more abundant at greater depths. The latter are known to behave as opportunistic species when great quantities of organic matter are available, and are usually associated to upwelling phenomena [19] [20]. Since the low level of hydrodynamic activity that allows deposition of fresh organic matter is also responsible for the settling of fine (silt and clay) particles, these foraminiferal species tend to be associated with fine deposits. Multivariate statistical analysis confirms the distributional trends of foraminiferal species mentioned above (Figure 6). *Group 2* appears closely related with fine sediments, whilst species of *Group 1* show negative correlation with the latter.

Differences between foraminiferal assemblages from the continental shelf north of the Nazaré Canyon and from the canyon itself (slopes) are probably related with the specific oceanographic and depositional conditions that characterized the latter. Benthic foraminiferal assemblages of the canyon are clearly less diverse than those of the adjacent shelf, probably reflecting higher instability of the canyon environment due to specific currents, strong turbulence and usually very high sediment supply, particularly in the canyon axis. In a study of live foraminferal assemblages from the Nazaré Canyon, [21] also related the very low total standing stocks (TSS) of benthic foraminifera in the upper and middle canyon axis to strong physical disturbance, leading to an unstable sediment substrate.

Species of the *Bolivinidae* family, identified by several authors (e.g. [22]) as microfaunistic proxies for upwelling occurrence, are more abundant in the canyon than on the continental shelf. This could be related to a higher intensity of this oceanographic phenomenon in the submarine canyon, caused by preferential passage of certain oceanic currents along the canyon. On the other hand, the species *Epistominella exigua*, which is known for fast development when large quantities of labile organic matter are available and therefore considered opportunistic under upwelling conditions, is common in the shelf but practically absent in the submarine canyon. This may be reflecting a *post-mortem* effect of the canyon's enhanced hydrodynamic activity on the settling and preservation of light-shelled and delicate foraminifera such as *Epistominella exigua*.

5. Conclusions

Although "gouf" type canyons act as preferential conduits for oceanic water masses towards the near-coast domain, the fossil record obtained from calcareous nannoplankton revealed that the higher diversity and abundance of nannoliths is not in the canyon itself, but on the northern adjacent shelf, away from the area influenced by canyon processes. This suggests that the Nazaré Canyon's complex and highly active hydrodynamic conditions are preventing nannoliths deposition and preservation inside the canyon, although locally settling and proliferation may be favoured.

In general, nannoliths show no significant correlation with TOC and silt-clay fraction suggesting that the majority of the studied samples record a (palaeo)ecological, rather than a taphonomical signature. The dominance of the smaller species in about 85% of the samples is clearly indicative of the good quality of the assemblages in this sector of the Iberian margin.

High abundance and diversity of nannoliths in energetic sectors of the inner shelf (sediments with low values of TOC and low content of silt-clay fraction) may be reflecting the occurrence of a persistent physical phenomenon possibly related with the canyon's morphology and proximity to the coast, *e.g.* injection of nutrients from the lower towards the upper levels of the water column, promoted by solitary internal waves generated at the canyon's northern shelf break. This hypothesis is confirmed by recent (unpublished) phytoplankton pigment data.

Benthic foraminiferal species reveal a strong correlation with the sedimentary cover in both the northern continental shelf and the Nazaré Canyon. Two groups of species were identified with different substratum preferences: *Group 1* develops in shallower environments of the shelf and submarine canyon, where coarser (sand and gravel) sediments prevail, probably due to high level of hydrodynamic activity that prevents the settling of fine material; *Group 2*, composed of species known to be opportunistic under high productivity conditions (*e.g.* upwelling), prefers colonization of fine sediments, usually present in deeper and less energetic environments. Lower foraminiferal content and lower diversity of species in the submarine canyon is probably due to its particular oceanographic and depositional conditions, associated with the geomorphologic setting and specific current patterns. The higher abundance of *Bolivinidae* (associated with higher organic matter content) in the canyon seems to reflect more intense upwelling phenomena here than on the adjacent shelf. The specific oceanographic settings of the canyon may also be conditioning the taphonomic record of benthic foraminifera. Strong currents and associated high hydrodynamic levels in the canyon seem to be preventing the preservation/settling of more fragile foraminifera shells (*e.g. Epistominella exigua*, an upwelling related species that is abundant on the northern shelf but almost absent in samples collected inside the canyon).

Preliminary results suggest that the Nazaré Canyon's peculiar hydro-sedimentary dynamic regime disturb benthic foraminifera productivity and diversity, or their preservation in the fossil record; and prevent settling of nannoliths in the more hydrodynamically active areas of the canyon, by generating hydrodynamic conditions similar to those of a typical inner shelf. Additionally, prevailing southward upwelling currents may have an important role in the distribution pattern of both calcareous nannoplankton and benthic foraminifera. This supports the hypothesis that the Nazaré Canyon influences the distribution pattern of surface thanatocenoses of Holocene, and probably older microfossils. Results will be developed in the near future, through the study of downcore distribution patterns in sediment cores collected from key-sectors of the Nazaré Canyon and its surroundings, and calibration with phytoplankton results currently under study at the Institute of Oceanography.

Acknowledgments
The authors wish to express thanks to João Vitorino (IH) and Rafael Mendes (IO) for their contributions and generosity during the discussion of the data. The text further benefited from extensive review from Henko de Stigter and João Alveirinho Dias. This work was funded by the scientific projects DEEPCO (POCTI/CTA/46367/2002), EUROSTRATAFORM (contract EVK3-CT-2002-00079 of the European Commission) and HERMES (contract of the European Commission).

References
[1] Arzola, R., Wynn, R., Lastras, G., Masson, D. & Weaver, P. (2008) – Sedimentary features and processes in the Nazaré and Setúbal submarine canyons, wst Iberian margin. *Marine Geology* (250): pp. 64 – 88
[2] Duarte, J., Dias, J.A. & Taborda, R. (2000) – Cabeceira do canhão da Nazaré: erosão versus sedimentação. *3º Simpósio sobre a Margem Ibérica Atlântica*, Faro 25-27: pp. 227 – 228.
[3] De Stigter, H., Boer, W., Mendes, P., Jesus, C., Thomsen, L., Van den Bergh, G., & Van Weering, T. (2007) – Recent sediment ytansport and deposition in the Nazaré Canyon, Portuguese continental margin. *Marine Geology* (246): pp.144 – 164
[4] Van Weering, T., de Stigter, H., Boer, W. & de Haas, H. (2002) – Recent sediment transport and accumulation on the NW Iberian margin. Progress in Oceanography (52): pp. 349 – 371
[5] Oliveira, A., Santos, A.I., Rodrigues, A. & Vitorino, J. (2007) – Sedimentary particle distribution and dynamics on the Nazaré canyon system and adjacent shelf (Portugal). *Marine Geology* 246 (2007) 105–122
[6] Duarte, F. (2002) – Distribuição espacial dos sedimentos da Nazaré e plataforma adjacente. *Proceed. 3.ª Asamblea Hispano-Portuguesa de Geodesia y Geofísica*, Espanha, 477 pp.
[7] Quaresma, L.S., Vitorino, J., Oliveira, A. & Silva, J. (2007) – Evidence of sediment resuspension by nonlinear internal waves on the western Portuguese mid shelf. *Marine Geology* (246): pp. 123 – 143
[8] Schmidt, S., de Stigter, H. & Van Weering, T. (2001) – Enhanced short-term sediment deposition within the Nazaré Canyon, North-East Atlantic. *Marine Geology* (173): pp. 55 – 67
[9] Flores, J.A. & Sierro, F.J. (1997) – Revised technique for calculation of calcareous nannofossil accumulations rates. *Micropaleontology*, 43 (3): pp. 321 – 324
[10] Pombo, J. (2004) – Sedimentos superficiais da plataforma continental portuguesa entre o Cabo Mondego e S. Martinho do Porto. *MSc thesis*, Faculdade de Ciências da Univ. de Coimbra, 185pp (unpublished)
[11] Loeblich, A. & Tappan, H. (1988) – Foraminiferal genera and their classification. *Van Nostrand Reinhold*, New York, 970pp
[12] Guerreiro, C., Cachão, M. & Drago, T. (2005) – Calcareous nannoplankton as a tracer of the

marine influence on the NW coast of Portugal, over the last 14000 years. *J. Nannoplankton Res.* 27(2): pp.159 – 172.

[13] McIntyre & Bé (1967) – Modern Coccolithophoridae of the Atlantic Ocean – I. Placoliths and Cyrtolyths. *Deep-Sea Research*, 14: pp. 561 – 597.

[14] Mendes, CR., Sá, C., Vitorino, J., Borges, C. & Brotas, V. (*in prep*) - Spatial distribution of phytoplankton communities in the Nazaré submarine canyon (Portugal)

[15] Cachão, M. & Moita, T. (2000) – *Coccolithus pelagicus*, a productivity proxy related to moderate fronts off Western Iberia. *Marine Micropaleontology*, 39 (1/4): pp. 131 – 155

[16] Brand, L. (1994) – Physiological ecology of marine coccolithophores. In: A. Winter, A and W. Siesser (eds). *Coccolithophores*, Cambridge University Press, Cambridge, pp. 39 – 149.

[17] Hickey, B.M. (1995) - Coastal Submarine Canyons. In: Proceedings 'Aha Huliko'a Hawaiian Winter Workshop 1995, *Topographic Effects in the Ocean,* pp. 95-110

[18] Levy, A., Mathieu, R., Poignant, A., Rosste-Moulinier, M., Ubaldo, M. & Lebreiro, S. (1995) – Foraminifères actuels de la Marge Continentale Portugaise: inventaire et distribution. *Memórias do Instituto Geológico e Mineiro*, nº32, 116 p.

[19] Mathieu, R. (1986) - Sediments et Foraminifères actuels de la marge continentale Atlantique du Maroc. *Thèse de Doctorat d' Etat ès Sciencies Naturelles.* Université Pierre et Marie Curie, 420 pp.

[20] Okada, H. (1983) – Modern nannofossil assemblages in sediments of coastal and marginal seas along the Western Pacific Ocean. In J.E. Meulenkamp (Ed.), *Reconstruction of Marine Paleoenvironments.* Utrecht Micropl. Bull.(30): pp. 171 – 187

[21] Koho, K., Kouwenhoven, T., de Stigter, H., van der Zwaan, G. (2007) – Benthic foraminifera in the Nazaré Canyon, Portuguese continental margin : Sedimentary environments and disturbance. *Marine Micropaleontology* (66): pp. 27 – 51

IOP Conf. Series: Earth and Environmental Science **5** (2009) 012005 doi:10.1088/1755-1307/5/1/012005

Elemental and stable isotopic approaches for studying the organic and inorganic carbon components in natural samples

J-F Hélie

GEOTOP-UQAM, C.P. 8888 succursale Centre-ville, Montréal (Québec), Canada, H3C 3P8

helie.jean-francois@uqam.ca

Abstract. The carbon cycle is an important part of major biogeochemical cycles. Many techniques may be used to characterize carbon amounts and sources in the environment. Here we first review the most popular techniques for the determination of organic and inorganic carbon concentrations. Decarbonatation techniques are also reviewed in details since it is often an important part of organic carbon analysis. The second part of this paper addresses the use of carbon stable isotopes to characterize organic carbon sources and processes in the environment. An overview of general stable isotopes background and terminology is given as well as the most popular analytical techniques.

1. Introduction

The carbon cycle has been the focus of many studies over the last few decades mainly because it is a very important part of biogeochemical cycles, but also because it is linked to climate change processes. However, fluxes between the different reservoirs of carbon are poorly understood, and especially carbon source identification. Organic carbon to nitrogen ratios (molar C:N ratios) can, and have been used to partition terrestrial vs. aquatically produced organic matter (OM). Terrestrially derived OM presents molar C:N ratios above 20, whereas aquatic organisms produce OM with molar C:N ratios between 4 and 10 [1]. However, bacteria are known to lower the C:N ratios of OM in soils by mineralizing nitrogen and respiring carbon as CO_2 (see [2] for a review). Hence, terrestrially derived OM loses part of its high C:N imprint when degraded by bacteria. As a result, this method usually has to be paired with another one to efficiently identify carbon sources. Nevertheless, different methods are used to measure organic carbon concentrations. Here we will review and describe the most popular techniques. Organic carbon stable isotopes have also proven to be very effective natural tracers of the organic carbon cycle in terrestrial and aquatic environments. Since stable isotopes are traditionally used less frequently than other methods, we will review the basic terminology, methodologies and applications for organic carbon geochemistry.

2. Carbon concentration determination techniques

In natural environments, carbon can be divided into its inorganic and organic forms. Inorganic carbon is the sum of aqueous CO_2 (mostly under the form of the weak carbonic acid H_2CO_3), bicarbonate (HCO_3^-), and carbonate (CO_3^{2-}) ions and their combination with different cations ($CaCO_3$, $MgCO_3$, etc.). Carbon comprised in molecules coming from living and once-living organisms is referred to as organic carbon. Carbon concentration determination techniques are different for inorganic vs. organic carbon materials.

2.1. Organic carbon concentration determination

Many methods are available to determine the bulk organic carbon concentration in natural materials. The most popular ones are the Loss on Ignition, Wet Oxidation and High Temperature Catalytic Combustion analysis.

2.1.1. *Loss on ignition.* This method is the least expensive and simplest of all organic matter concentration determination methods. The dry, ground and homogenized sample is weighed and transferred to a clean ceramic cup. The ceramic cup is then placed in a high temperature oven and heated to 500-550°C for about 2 hours. The sample is then weighed again and the amount of OM is given by the weight difference (pre-combustion – post-combustion). Since all of the OM is combusted, with no separation of the resulting gases, the results are expressed in mg of OM rather than in mg of organic carbon. Although one might apply a simple correction factor to estimate the amount of organic carbon from total OM content, this factor, which is in fact the average percentage of carbon in OM, actually varies from one sample to the next and even more so from one type of material to another. Moreover, some samples may present important amounts of clay-bound water (up to 20% of the total weight) that vaporize during combustion. For such samples, the estimation of organic carbon concentration would be that much more erroneous. Great care should thus be taken when interpreting organic carbon content from loss on ignition.

2.1.2. *Wet oxidation.* This method is mostly used to provide organic carbon content in water samples and requires an analyzer. Different companies use different techniques that all rely on the same principle. An aliquot of the sample water is transferred to a chamber where sodium persulfate is added at a temperature of about 100°C. Carbon from the sample OM is then oxidized to CO_2 and transferred, through a carrier gas (usually nitrogen or helium) to a detector. The amount of CO_2 detected is converted to an organic carbon concentration through a calibration curve performed regularly (every day, week or month depending on applications and\or lab practices) with known concentrations of organic carbon. Inorganic carbon is not oxidized and does not interfere with organic carbon concentration measurements. The organic carbon analyzer does not discriminate between the dissolved and particulate phases. The sample thus has to be filtered before analysis, preferably immediately following sampling since the dissolved and particulate phases can exchange matter through the colloid phase. Although the operationally-defined limit between the dissolved and particulate phases is 0.45 µm, one can choose from a range of filter porosity (from 0.2 to 1.0 µm) mostly because glass fibre filters manufacturers do not offer 0.45 µm porosity filters.

2.1.3. *High Temperature Catalytic Combustion analysis.* This method is probably the most accurate of all. However, since the combustion occurs at a very high temperature and all forms of carbon are converted to CO_2, one must take great care in selecting a suitable decarbonatation method to efficiently remove inorganic carbon while preserving organic carbon. There are two main decarbonatation methods, namely *i)* acidification in solution, and *ii)* acidification by fumigation.

i) For the acidification in solution method, a sufficient amount of ground and dried sample (~50 mg) is placed in a beaker. Hydrochloric acid (HCl 1N) is added to the sample and left to react for 24 hours. The sample is then either left to evaporate until completely dry or the excess acid is gently decanted (great care must be taken not to pour off any of the sample) and replaced with distilled water (and decanted until neutral pH is attained). The dry sample is then carefully weighed and analyzed. This method insures complete decarbonatation. However, acid-soluble OM may be poured off with excess acid and/or distilled water thus underestimating organic carbon content during analysis. Worth of mention here is also the fact that the potential loss of acid-soluble OM may induce fractionation of carbon and nitrogen stable isotopes, thus leading to erroneous $\delta^{13}C$ and $\delta^{15}N$ measurements on the bulk sample. Also, one needs to correct the % weight results for the loss of carbonate material since we need % results with respect to the total weight

IOP Conf. Series: Earth and Environmental Science **5** (2009) 012005 doi:10.1088/1755-1307/5/1/012005

of the sample (including carbonate material). To take into account the loss of carbonates, one needs to also analyze the total carbon of the sample (i.e. without decarbonatation) and apply the following calculation:

$$C_{org} = 100 \times R_2 \times \left\{ \frac{[1-(R_2-R_1)]}{(R_2-0.12)} \right\} \qquad \text{Equation 1}$$

Where C_{org} is the weight percentage of organic carbon in the sample corrected for loss of carbonates, R_2 is the weigh percentage of carbon after acid treatment (result of acid treated sample analysis ÷ 100) and R_1 is the weight percentage of total carbon (result of untreated samples analysis ÷ 100).

ii) For the fumigation method, the sample is carefully weighed into a silver cup (used for High Temperature Catalytic Combustion analysis) and placed in a clean tray. The tray is then placed in a closed glass container with a beaker of concentrated HCl for 24 hours. The air in the glass container rapidly saturates with HCl which reacts with the carbonate minerals in the sample. The silver cup is then closed and analyzed with an EA. Since the sample was weighed before decarbonatation, the result is given in % of total weight and no correction is needed. Tin, although it is a better catalyst for analysis, cannot be used for fumigation because it reacts with the acid and breaks up. However, it is recommended to wrap the silver cup in a tin cup after fumigation. Also, since silver is not as good a catalyst as tin, its combustion is less complete and an insert needs to be used in the reaction tube of the EA to remove ashes more often, making this method a little more expensive than the other one. Moreover, one must not use any plastic material inside the fumigation chamber (including the tray), as certain types of plastics and other organic polymers may react with acid, vaporize and deposit onto the sample, thus overestimating organic carbon concentrations. Glass, Teflon and non-reactive metals such as titanium are more suitable for plate material as they do not react with acid. Teflon is an expensive material but drilling is necessary to accommodate the silver cups making glass very difficult to use.

The principle of simultaneous analysis of carbon, hydrogen, nitrogen and sulphur analysis (High Temperature Catalytic Combustion) is relatively simple and based on the principle of the Liebig reaction (1930 – according to [3]). The sample, wrapped in a catalyst (usually tin but silver is also used) is dropped from a sequential carousel into an oven simultaneously with O_2 and combusted at high temperature (1000°C). The resulting gases are transferred to a reducing environment (copper wires, turnings or beads heated to 800°C) through a carrier gas (usually helium) to reduce nitrogen oxides (based on the Dumas method – 1831 – according to [3]) and sulphur oxides to N_2 and SO_2, respectively. The final gas mixture is then sent to a gas separating device (usually a gas chromatography column or GC column, but a chemical trap using a temperature program desorption trap is also used) to allow determination of the amount of individual gases on a detector. The detector is usually a thermal conductivity detector (TCD) and amounts of the different gases are deduced from a calibration curve. For CNS analysis, it is recommended to use a water trap (magnesium permanganate, dryrite or phosphorus pentoxide) before gas separation, especially if the elemental analyzer (or EA) is connected to a mass spectrometer to determine the isotopic composition of the resulting gases. Figure 1 illustrates the usual setup for High Temperature Catalytic Combustion analysis using a conventional Elemental Analyzer (EA).

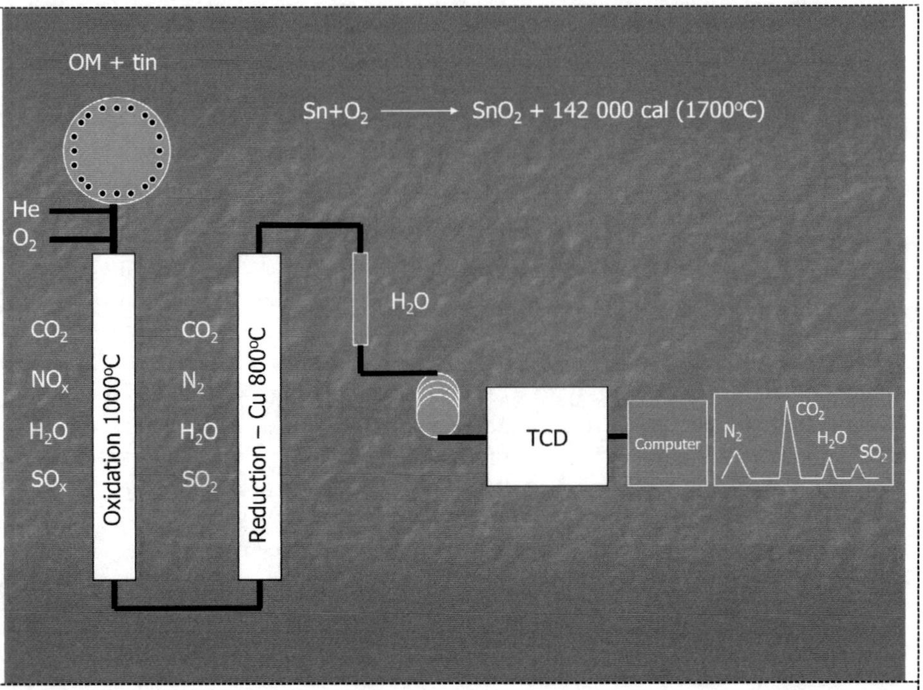

Figure 1. Typical High Temperature Catalytic Combustion analysis setup sequence using a conventional EA. Note that the use of a tin catalyst combined with the injection of O_2 at 1000°C produces a "flash" that locally raises the temperature to about 1700°C. Also note that if a water trap is used, no determination of H_2O is possible on the gas chromatogram.

2.2. Inorganic carbon concentration determination

Inorganic carbon concentration can either be obtained by measuring directly the amount of inorganic carbon by calcimetry (inaccurate) or coulometry (very accurate), or indirectly from total carbon and organic carbon measurements.

2.2.1. *Coulometry.* Coulometric titration is an absolute and accurate method of determining CO_2 amounts in a stream of gas. For inorganic carbon concentration determination, the sample is weighed in a Teflon or ceramic cup and placed in a reaction tube. The reaction tube is attached to the acidification unit. Perchloric acid ($HClO_4$ 2N) is dropped on the sample and CO_2 is released from the carbonate material (mostly $CaCO_3$ & $MgCO_3$) into a stream of N_2 gas (carrier gas) that is delivered to the titration unit. In the titration unit, there is a titration cell, containing a cathode, an anode, a solution of monoethanolamide, a colorimetric pH indicator and a photo detector. When CO_2 enters the cell, it is quantitatively absorbed by monoethanolamide to form a titratable acid. The added acid causes the colorimetric pH indicator to change colour which is detected by the photodetector as a percentage of transmittance change. A titration current is triggered to electrochemically produce a base proportionally to the % transmittance change. The result is an absolute amount of CO_2 from the principles of Faraday's Law (1 faraday of electricity will result in the alteration of 1gram equivalent weight or GEW of a substance during electrolysis). Each faraday of electricity expended is equivalent to 1 GEW of CO_2 titrated. Even though this method is absolute and does not require the use of a calibration curve, it is recommended to intersperse known standards in the analytical sequence to verify the performance of the instrument over time. Athough sulphuric (H_2SO_4) and hydrochloric

(HCl) acids may be used to acidify the sample, dilute perchloric acid (2N) is preferred since H_2SO_4 may occlude some of the inorganic carbon and HCl can overload the instrument's scrubber.

2.2.2. *Calculation from total carbon & organic carbon.* Since the EA gives total carbon concentration, it is relatively straightforward to obtain inorganic carbon concentration from first, the analysis of total carbon and second, the analysis of organic carbon after decarbonatation of an aliquot of the same sample. However, if weighing of the sample occurs before decarbonatation (i.e. acidification in solution), one needs to take into account the loss of inorganic carbon before reporting the results to % weight (see section 2.1.3). Then, inorganic carbon concentration is given by the subtraction of organic carbon concentration to total carbon concentration. The use of the calculation method is less precise than coulometry since the precision is obtained from the propagation of the error on each measurement.

3. Organic carbon isotopic composition determination techniques

3.1. Stable isotopes: background and terminology

What defines an element is the number of protons contained in the atom core, and the mass of an element is given by the sum of its protons and neutrons. For example, the core of every atom of carbon contains 6 protons and if the core of a given atom of carbon contains 6 neutrons, then its mass is 12 AMU (atomic mass units). Isotopes of a given element have the same number of protons (what defines an element) but have a different number of neutrons. Thus, carbon atoms that contain 7 neutrons have a mass of 13 and carbon atoms that contain 8 neutrons have a mass of 14 (respectively ^{13}C and ^{14}C). For some isotopes, the combination of protons and neutrons in the core is not stable and the core will lose neutrons or protons or both to produce another isotope or another element of lower mass and\or atomic number. These unstable isotopes are thus radioactive and disintegrate following an exponential function. Some isotopes have a stable combination of protons and protons and are thus called stable isotopes. Stable isotopes of a given element all participate in the same biogeochemical reactions. However, since they do not have the same atomic mass, the energy required in those reactions is different (i.e. more energy is required for heavier atoms) and their distribution in nature varies accordingly. Stable isotopes are thus very often effective natural tracers of sources and pathways in natural environments.

To measure stable isotopes in the different natural materials, one needs to use an isotope ratio mass spectrometer (see section 3.2). The very nature of those instruments dictates the manner we report stable isotopic measurements. Since ionization of the sample gas in the source produces large fractionation between the different masses, the stable isotopic community has very early on decided to report results with respect to international reference materials. The reference material is different for almost every element analyzed (table 1).

Table 1. Most common international reference materials used to report stable isotopes data

Reporting international reference material	Reported in text as	Element	Material
Vienna Pee Dee Belemnitella	V-PDB	Carbon & Oxygen	Carbonates
Vienna Standard Mean Ocean Water	V-SMOW	Oxygen & Hydrogen	Water
Atmospheric air N_2	AIR	Nitrogen	All
Vienna Cañon Diablo Troilite (meteorite)	V-CDT	Sulphur	All

Results are thus reported as the ratio of the amount of rare isotope to the amount of abundant isotope in the sample with respect to the international reference material as a delta (δ) unit (equation 2). And since these ratios are very small, we usually multiply delta units by a thousand and report results in permil (‰).

$$\delta^{13}C = \left(\frac{N_S(^{13}C)/N_S(^{12}C)}{N_{ref}(^{13}C)/N_{ref}(^{12}C)} - 1\right) \quad \text{Equation 2}$$

Where N_s is the amount of the element in the sample and N_{ref} is the amount of the element in the reference material.

3.2. Stable isotopes analysis

The mass spectrometer was invented in the late 1800s, early 1900s. The mass spectrometer was then referred to as *parabola spectrograph* and mass spectrometry was referred to as *positive ray analysis* [4] & [5]. Since then, mass spectrometers have evolved considerably [6]. However, the basic components are still an ionizing source, an analyzer (in this case a magnetic sector) and detectors (Figure 2). For light stable isotopes, we use a gas source and different preparation systems are now used to convert the original material in a gas that can be introduced in the mass spectrometer.

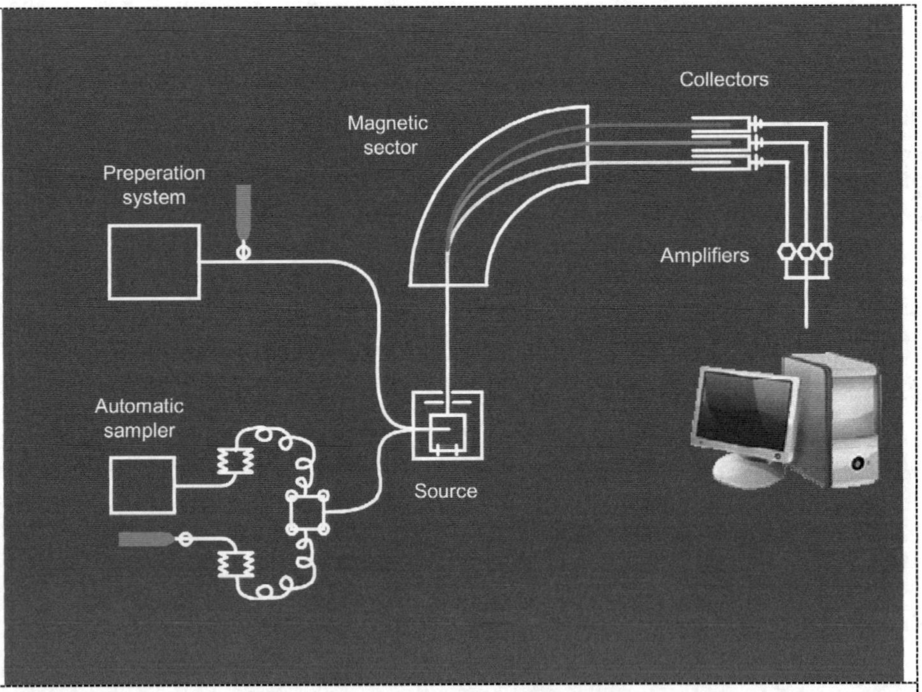

Figure 2. Typical stable isotopes ratio mass spectrometer components in both dual inlet and continuous flow configurations.

In the case of isotopic analysis of bulk organic carbon, the mass spectrometer is coupled to an EA (see section 2.1.3.). The dry, ground, homogenized and decarbonated sample is weighed in a tin cup. The actual weight is not very important; however, to avoid any non-linearity issues in the mass spectrometer, the amount of gas generated by the sample has to remain constant and be as close as possible to the amount of monitoring or reference gas injected during analysis. The sample is dropped via an autosampler into the EA where it undergoes oxidation, reduction and gas separation (see section 2.1.3.) before entering the mass spectrometer. Gases analyzed by the standalone EA are usually vented to the laboratory atmosphere. In the case of EA-IRMS however, a stainless steel tube is attached to the vent to carry the gases to the IRMS. A silica tube is inserted in the stainless steel tube to create an open split (or gas sampling device depending on the manufacturer). The silica tube is then inserted into the inlet of the mass spectrometer. Gases produced by the EA are thus carried to the mass spectrometer in a helium gas stream and excess pressure is relieved through the open split. CO_2 enters the mass spectrometer source where it is ionized by a stream of electrons generated by a thorium filament. The ionized gas is accelerated using a potential difference and focused with electromagnetic lenses to produce an optimized beam of ionized CO_2 in a flight tube under vacuum. The beam is then deflected in a magnetic field generated by the electromagnet. This results in the separation of the different masses of CO_2 (or of the studied gas – N_2, SO_2, CO, etc.). The separated mass beams are collected in Faraday cups, where the intensity of the measured current is proportional to the amount of each mass collected. Since the measured currents are very small, they are amplified before being sent to the acquisition software of a computer. Most commercial mass spectrometers are sold with universal triple collector configuration (or Faraday cups). In the case of CO_2, the collectors measure masses 44, 45 and 46 (Table 2).

Table 2. Combinations of carbon and oxygen isotopes for the different masses of CO_2

Mass measured in the Faraday cup	Combinations of isotopes in CO_2
44	$^{12}C\ ^{16}O\ ^{16}O$
45	$^{13}C\ ^{16}O\ ^{16}O$
	$^{12}C\ ^{17}O\ ^{16}O$
46	$^{12}C\ ^{18}O\ ^{16}O$
	$^{13}C\ ^{17}O\ ^{16}O$

Since only a fraction of the ionized CO_2 is collected by the detectors (about 50%), fractionation occurs in the ion source and absolute ion counting is not possible, hence the use of the delta notation to report results (equation 2; section 3.1). In practice, every sample gas is preceded or followed by either a reference CO_2 (dual inlet mode) or monitoring CO_2 (continuous flow mode) that is analyzed under the same conditions as the sample gas. Monitoring gas is supplied by a large gas cylinder as compressed CO_2. Through the course of a day, the isotopic composition of such a gas does not change significantly. Since it does change on a time scale of weeks to months, any analytical sequences should contain a number of reference materials of known isotopic compositions. Usually, 2 or 3 different reference materials are measured periodically in the analytical sequence (once every 10 samples for example). Using the reference materials analyzed during an analytical sequence, the measured delta values of the samples are corrected to "true" delta values. International reference materials are available for purchase [7]. However, since they are sold in limited quantities and at a high price it is recommended to find internal laboratory reference materials and calibrate them using the international reference material. It is important to select internal reference materials that are homogenous, that will not change over time and will bracket the expected values of the samples in an analytical sequence.

Analytical uncertainty is usually estimated by analyzing 10 replicates of a sample. Analytical uncertainty should be repeated for different types of samples and does not take into account the representiveness of the sample.

4. Synthesis and concluding remarks

Carbon, in natural environments, is divided into its inorganic and organic forms. Although organic carbon may be determined through different methods, it is most efficiently and accurately done through High Temperature Catalytic Combustion analysis coupled to an efficient decarbonatation method. Similarly, inorganic carbon determination is most accurately achieved through coulometry using dilute perchloric acid.

Stable carbon isotopes have proven to be effective tracers of the carbon cycle in natural environments. One must take great care in reporting stable carbon isotopes data in the correct nomenclature to avoid confusion and report data to the correct international reference material. Organic carbon isotopic composition determination is efficiently achieved using a universal triple collector gas source isotope ratio mass spectrometer coupled to an elemental analyzer in continuous flow mode. An efficient decarbonatation method must be used prior to analysis and data must be corrected using known in house reference material interspersed in every analytical sequence.

Acknowledgments
The author would like to thank Guy Bilodeau and Yves Gelinas for their very helpful comments.

References

[1] Meyers PA 1994 Preservation of elemental and isotopic source identification of sedimentary organic matter *Chem. Geol.* **114** 289-302
[2] Hélie J-F and Hillaire-Marcel C 2006 Sources of particulate and dissolved organic carbon in the St. Lawrence River: isotopic approach Hydrol. Proc. **20** 1945-1959
[3] Pansu M and Gautheyrou J 2006 Organic and Total C, N (H, O, S) Analysis *Handbook of Soil Analysis* (Springer) pp 329-370
[4] Thompson J J 1897 Cathode Rays *Phil. Mag.* **44** 293
[5] Thompson J J 1913 Rays of positive electricity *Proceedings of the Royal Society* A 89 1-20
[6] Brand W 2004 Mass Spectrometer Hardware for Analyzing Stable Isotope Ratios *HandBook of Stable Isotope Analytical Techniques* Vol.1 (Elsevier) ed PA de Groot pp 835-856
[7] Coplen TB, Hopple JA, Böhlke JK, Peiser HS, Rieder SE, Krouse HR, Rosman KJR, Ding T, Vocke RD, Révész KM, Lamberty A, Taylor P and De Bièvre P 2001 Compilation of minimum and maximum isotope ratios of selected elements in naturally occurring terrestrial materials and reagents *USGS Water-Resources Inversigation Report 01-4222* 98p

IOP Conf. Series: Earth and Environmental Science **5** (2009) 012006 doi:10.1088/1755-1307/5/1/012006

Compound-specific radiocarbon analysis – analytical challenges and applications

Gesine Mollenhauer and Janet Rethemeyer

Alfred-Wegener Institute for Polar and Marine Research
Am Handelshafen 12, 27570 Bremerhaven, Germany

E-mail: Gesine.Mollenhauer@awi.de

Abstract. Within the last decades, techniques have become available that allow measurement of isotopic compositions of individual organic compounds (compound-specific isotope measurements). Most often the carbon isotopic composition of these compounds is studied, including stable carbon ($\delta^{13}C$) and radiocarbon ($\Delta^{14}C$) measurements. While compound-specific stable carbon isotope measurements are fairly simple, and well-established techniques are widely available, radiocarbon analysis of specific organic compounds is a more challenging method. Analytical challenges include difficulty obtaining adequate quantities of sample, tedious and complicated laboratory separations, the lack of authentic standards for measuring realistic processing blanks, and large uncertainties in values of $\Delta^{14}C$ at small sample sizes. The challenges associated with sample preparation for compound-specific $\Delta^{14}C$ measurements will be discussed in this contribution.

Several years of compound-specific radiocarbon analysis have revealed that in most natural samples, purified organic compounds consist of heterogeneous mixtures of the same compound. These mixtures could derive from multiple sources, each having a different initial reservoir age but mixed in the same terminal reservoir, from a single source but mixed after deposition, or from a prokaryotic organism using variable carbon sources including mobilization of ancient carbon. These processes not only represent challenges to the interpretation of compound-specific radiocarbon data, but provide unique tools for the understanding of biogeochemical and sedimentological processes influencing the preserved organic geochemical records in marine sediments. We will discuss some examples where compound-specific radiocarbon analysis has provided new insights for the understanding of carbon source utilization and carbon cycling.

1. Introduction

After its discovery in the first half of the 20th century, the method of radiocarbon dating became widely used in a number of disciplines, including archaeology and the geosciences. The method relies on measuring the radiocarbon (^{14}C) content of a given sample composed of either carbonate, bones or any organic material. Combined knowledge of the initially present level of ^{14}C and its radioactive halflife allows the calculation of a ^{14}C-age of the sample. In the simplest case of a sample composed homogenously of a material that was synthesized using a well constrained carbon source like atmospheric carbon dioxide, this method allows accurate determination of the age of the sample.

Most organic samples used for radiocarbon dating in the geosciences contain a mixture of organic substances, each with a different radiocarbon signature [1]. This is particularly true for total organic carbon in sediment and soil samples, but may also apply for biomass of living organisms that incorporate carbon from several sources, i.e., heterotrophic organisms (cf. [2]). Thus, radiocarbon ages of total organic carbon samples are complicated to interpret and in most cases cannot be directly taken for the determination of the sample's age. Using total organic carbon for dating purposes may result in misleading interpretations [3, 4].

The problem of heterogeneous composition of organic materials from unknown sources and with unknown relative contributions can be circumvented when focusing on source-specific organic compounds, so-called biomarkers and analysing their radiocarbon content (compound-specific radiocarbon analysis (CSRA)). Typical biomarkers include lipids synthesized by marine planktonic algae (e.g., alkenones), compounds originating from epicuticular wax layers of land plants (e.g., long-chain *n*-alkanes or *n*-fatty acids), or membrane lipids from archaea. If these compounds can be purified in sufficient quantities from heterogeneous mixed samples to allow for measurement of their radiocarbon content, more specific information can be gained on processes governing their formation.

The method of CSRA was introduced and first results of individual *n*-alkanes and *n*-fatty acids were published in 1996 by Eglinton and co-workers [5]. Since then, CSRA has been extended to a number of organic compounds [3, 6, 7], and sample size requirements have been reduced to at least 25 μg of carbon [8]). Major challenges in CSRA are (a) the purification of adequate quantities of individual compounds and (b) the conversion of the biomarker compounds into graphite cathodes for ^{14}C analysis by accelerator mass spectrometry (AMS) (c) without introducing contaminants. The methodical challenges associated with sample preparation for CSRA and recent examples of applications of CSRA in geochemical studies will be discussed in this contribution.

2. Analysing compound-specific radiocarbon contents

2.1 Sample purification

In natural samples most compounds serving as biomarkers are only present in relatively small concentrations (typically $<1\mu g\ g^{-1}$ dry weight). These compounds need to be extracted in high purity and in sufficient quantity to allow for radiocarbon analysis by AMS. A reasonable amount of biomarker to be aimed for is at least 100 μg of carbon. Therefore, typically large samples >100 g dry weight are required.

For extraction of compounds from natural samples, liquid extraction techniques are used most often. For lipids, for example, extractions with organic solvents are performed. Several equally adequate procedures exist [9, 10], including the use of ultrasonic probes [11], Soxhlet extraction [12], or accelerated solvent extraction (ASE) [9]. From the total lipid extract, individual compounds can be purified using a sequence of wet chemical techniques and chromatographic steps of separation. Depending on the target compounds, these steps may include acid or base hydrolysis (base hydrolosis is often referred to as "saponification"), and several chromatographic steps separating compound classes according to their polarity (e.g., silica gel (SiO_2) columns), degree of saturation ($AgNO_3$ coated silica gel), or ion exchange capacity [13]. Further steps may involve molecular sieving techniques like urea adduction [14], e.g., to separate straight chain from branched and cyclic compounds. Ohkouchi et al. [15] describe a dedicated protocol for isolation of long-chain unsaturated ketones, so-called alkenones, as a compound class. These compounds are of particular interest because of their well-defined phytoplanktonic source. Other possible compound classes isolated by a combination of several of the above mentioned techniques include *n*-alkanes [7], or total hydrolysable amino acids [13].

If purification of individual compounds from "pre-cleaned" compounds classes is desired, instrument-based chromatographic techniques must be employed. For compounds amenable to gas chromatography, preparative capillary gas chromatography can be used [5]. For this method, a gas chromatograph (GC) is interfaced with a programmable fraction collection (PFC) device. A splitter diverts 1% of the total eluate to a detector, while 99% reach the PFC. Glass traps attached to the PFC can be programmed to collect the compounds of interest based on their retention times. Repeated injection of the same sample ensures recovery of sufficient amounts of the target compound for radiocarbon analysis.

For more polar or higher molecular weight compounds, high performance liquid chromatography (HPLC) is an alternative method. HPLC can be interfaced with a fraction collector collecting the effluent at the time, when the compounds of interest elute from the column [16]. The collection windows can either be time-based [16], or based on detection of target masses in a mass spectrometer if a splitting device is used to monitor the eluate [3].

In either case, purity of the target compound or compound-class must be ensured prior to radiocarbon analysis. Sample purity can be monitored by analysing a small sub-sample of the purified compound by GC or HPLC. GC or HPLC chromatograms devoid of peaks representing contaminant compounds are a first indication for sample purity, however, contaminants outside of the respective analytical window cannot be excluded. An additional indication of sample purity is the comparison of the CO_2 yield after combustion in evacuated and flame-sealed quartz-glass tubes with the expected carbon amount based on compound quantification by GC or HPLC. GC- or HPLC-based quantification can additionally be checked by weighing of purified samples using a micro-balance.

2.2 AMS measurement of small samples

Purified samples to be analysed by AMS must first be converted to CO_2 (by combustion for organic materials or by acid hydrolysis for carbonates), which subsequently is reduced to graphite with an excess of hydrogen over iron or cobalt as catalyst [17, 18]. The graphite is pressed into aluminium target holders that are placed in the sample carousel of the AMS ion source. For standard AMS ^{14}C measurements, a minimum sample size of 1 mg carbon is required. However, techniques have been developed at several facilities that allow a drastic reduction of sample size. Very small sample sizes of less than 50 μg of carbon can now be analyzed at several AMS facilities that have developed special small volume reactors for sample CO_2 conversion into graphite cathodes leading to a considerable reduction of background, mainly isotopic fractionation, during this procedure (e.g., [17, 19, 20]). When using such systems, the blank associated with this step of the sample preparation procedure could be reduced to about 0.1 μg of carbon [20].

AMS ^{14}C measurements are performed by determining the ratio of ^{14}C and the stable carbon isotope (^{12}C or ^{13}C) of a sample in comparison to that of a standard (HOxII, NIST). The isotope ratio is calculated from the ^{12}C and/or ^{13}C ion currents detected in Faraday cups and the number of ^{14}C counts that are often measured in a gas ionization detector. In previous studies, AMS ^{14}C measurements of several subsamples of variable size taken from one larger sample of CO_2 showed a decrease in measured $^{14}C/^{12}C$ ratios below a certain sample size, i.e., measurement accuracy decreases with decreasing sample size [17, 20, 21]. This has been attributed to low ion currents produced from small samples in the AMS ion source and resulting mass dependent isotopic fractionation that was found to be different for the stable carbon isotope and for ^{14}C [17, 20]. This difficulty can be circumvent by preparing and analyzing standards of similar small sample size in the same measurement cycle, i.e. AMS sample carousel [17, 21]. The deviation of measured $^{14}C/^{12}C$ ratios of the small standards from the consensus value for the standard material can be quantified and used as a correction factor to be applied to the measured $^{14}C/^{12}C$ ratios of small unknown samples of similar size.

2.3 *Blank assessment*

Each step of the sample treatment procedure for AMS ^{14}C analysis, including, e.g., purification, combustion, and graphitization, may add small amounts of carbon from unknown sources and with unknown isotopic composition to the sample, so-called process blank. The assessment of blank carbon contribution is essential for interpretation of radiocarbon data, in particular if samples to be analyzed are small. Since most purified biomarker samples only contain 50 – 300 μg C, blank-assessment is particularly important for CSRA. Special care must be taken when ultra-microscale (<25μg C) measurements are made. While the measurements are technically feasible, the uncertainty of the results increases tremendously with decreasing sample size since the relative importance of the blank carbon addition increases with decreasing sample size. However, measuring size and isotopic composition of blanks associated with the entire sample treatment (procedure blank), or with individual steps of the sample preparation process (e.g., combustion blank, graphitization blank) is not trivial.

One possible approach is to analyze material of known isotopic composition. Commercially available reference materials are often sold including information on δ^{13}C, but ^{14}C concentration is not a standard property determined by manufacturers. Therefore, even when using a reference material, a Δ^{14}C value of this material has to be determined prior to processing the sample. Furthermore, it is advisable to use two types of reference materials, one with modern radiocarbon content and one without any measurable ^{14}C content (radiocarbon-"dead") in order to monitor possible modern or ^{14}C-dead carbon contributions.

In our laboratories in Bremen, we tested the blank contribution of the GC-PFC system by isolating a modern n-$C_{16:0}$ fatty acid standard material and its effect on the measured δ^{13}C and Δ^{14}C values (Figure 1). This procedure could be considered an assessment of a partial procedure blank. For this purpose, some of the standard material was dissolved in hexane, derivatized with methanol of known isotopic composition to form a fatty acid methyl ester, and injected approximately 65 times into the GC-PFC until ~ 300 μg of the compound could be recovered from the PFC glass trap. The trapped compound was washed into a vial with 1 ml dichloromethane, and eluted over a silica-gel column to remove possible column bleed. Consequently, the fatty acid methyl ester as well as 1.52 mg of the untreated fatty acid standard material were dissolved in dichloromethane and transferred in pre-combusted (900°C, 5 h) quartz glass tubes together with pre-combusted cupric oxide (CuO). The quartz-glass tubes were attached to a vacuum line, pre-evacuated to ~10^{-3} mbar, and further evacuated to a pressure of ~10^{-5} mbar while immersed in a slurry of dry ice and ethanol to avoid sublimation of purified compounds. Finally during flame-sealing, quartz-glass tubes were cooled in liquid nitrogen. This step was followed by combustion over night at 900°C. Resulting CO_2 was dried in a vacuum line equipped with a water trap cooled in a slurry of dry ice in ethanol, and submitted for δ^{13}C and radiocarbon analysis to the National Ocean Science Accelerator Mass Spectrometer (NOSAMS) facility in Woods Hole, USA. Results reported by NOSAMS for the fatty acid methyl ester sample had to be corrected for carbon contribution of the methyl group using isotopic mass balance calculation. Both measurements yielded similar values within or slightly outside the 1σ error range (Figure 1). However, the standard material processed by GC-PFC was slightly more depleted in heavy isotopes than the untreated standard material indicating that any blank contributed by the GC-PFC method is likely depleted in both ^{13}C and ^{14}C and therefore probably originates from fossil carbon, e.g., from "column bleed" or from the organic solvents used.

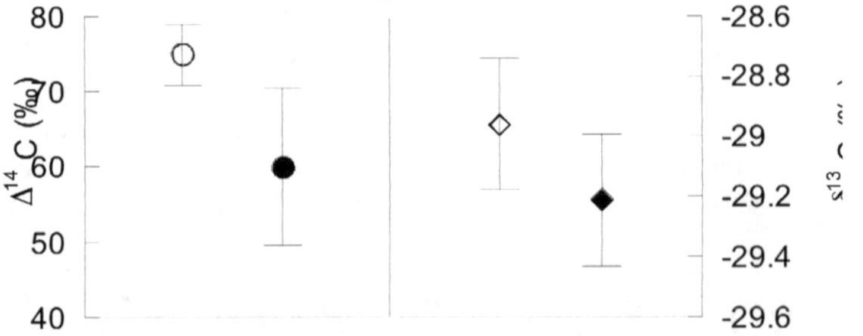

Figure 1: Carbon isotopic composition of n-$C_{16:0}$ fatty acid standard dissolved in dichloromethane (open symbols) and after purification using GC-PFC (closed symbols). Error bars are 1σ errors for Δ^{14}C, and average standard deviations of measurements for δ^{13}C.

A similar test was performed to check the GC-PFC system at the Max-Planck-Institute for Biogeochemistry in Jena, Germany, and subsequent conversion of isolated compounds into graphite for ^{14}C analysis at the AMS facility in Kiel, Germany [22]. Two fatty acid standards n-$C_{18:0}$ and n-$C_{28:0}$ were first radiocarbon dated without any treatment yielding 103.8 and -272.4 ‰, respectively. To test the effect of solvent evaporation on the ^{14}C content, several mg of each standard were dissolved in dichloromethane and the solvent was evaporated under a stream of nitrogen while heating at 50°C. ^{14}C concentrations for both standards were not significantly (1 σ range) different from that of the non-treated fatty acids (Figure 2). In contrast, ^{14}C values obtained for isolated standards using GC-PFC were depleted in ^{14}C slightly outside the 1 σ measurement uncertainty indicating the admixture of old or fossil carbon during compound isolation using GC-PFC and/or subsequent conversion into graphite.

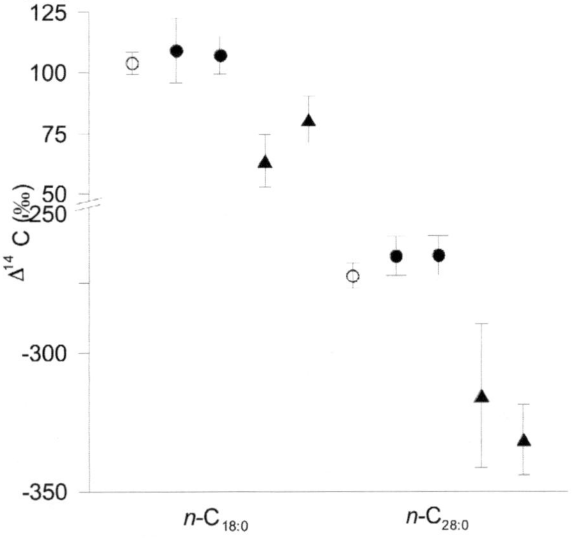

Figure 2: ^{14}C concentration of n-$C_{18:0}$ and n-$C_{28:0}$ fatty acid standards: untreated (open symbols), dissolved in dichloromethane (filled circles) that was evaporated subsequently, and isolated by GC-PFC (triangles) with 1 σ measurement uncertainty. Results for two separate purification runs are displayed.

In an earlier experiment, the entire purification procedure for alkenones, including combustion and graphitization, as described by Ohkouchi et al. [15] was examined for blank contribution, and the difference between small and standard sized samples was investigated [23]. For this purpose, algal cultures of the alkenone producer *Isochrysis galbana* were grown in two batches. One batch was kept in seawater in contact with the atmosphere and thus contained DIC with modern ^{14}C levels. The second batch was grown after replacing modern DIC with DIC strongly depleted in ^{14}C. In order to assess the blank contribution of the extraction and purification procedure, the assumption was made that alkenones should record ^{14}C levels of ambient DIC. In both experiments, $\Delta^{14}C$ values of several sub-samples of alkenones spanning a size range between 39 and 2203 μg C were analyzed. If a systematic addition of modern carbon would occur, the $\Delta^{14}C$ values of alkenones from the ^{14}C-depleted culture would be expected to deviate strongly from the DIC value. A fossil carbon blank would be expected to have a strong effect on the alkenones from the modern culture. If combined combustion and graphitzation blanks added significant amounts of carbon, the small samples should be affected more strongly than the larger samples. Results agreed in most cases with $\Delta^{14}C$ values of the ambient DIC [23]. Agreement between sub-samples was also acceptable, and no clear trend was observed with sample size. The authors reached the conclusion that no significant amounts of either modern or ^{14}C-depleted blank carbon are systematically introduced during the purification procedure. In our laboratories in Bremen, we recently re-purified one sub-sample of the alkenones from both cultures, and the new results fall exactly within the scatter of original data (Figure 3). These results imply that the purification method adds insignificant amounts of C to the samples.

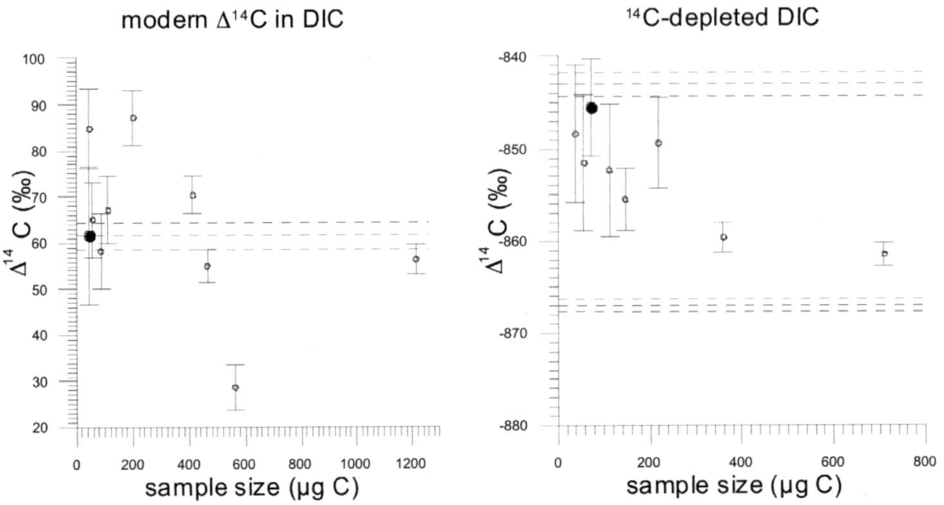

Figure 3: Compiled $\Delta^{14}C$ values of alkenones extracted from algal cultured grown in seawater with modern and ^{14}C-depleted DI^{14}C levels, shown with 1σ error bars. Closed symbols are samples purified in Bremen. Small open symbols denote data from Mollenhauer et al. [23], samples purified in Woods Hole. Dashed horizontal lines mark $\Delta^{14}C$ of DIC with 1σ errors, in the ^{14}C depleted culture the lower $\Delta^{14}C$ levels were measured before the culturing, the higher after the culturing. Smaller samples are slightly ^{14}C-enriched. Note the larger error bars for samples smaller than 300 μg C, for which the dedicated AMS techniques described in section 2.2 are applied.

A more direct way to determine blank contribution is to measure the amount of carbon yielded from a true process blank. In a recent study, the blank associated with HPLC based purification of archaeal biomarkers was determined by collecting various amounts of HPLC effluent, drying the solvent under N_2, and combusting the residue in pre-combusted quartz glass tubes with CuO [8]. CO_2 was measured manometrically, and subsequently converted to graphite for AMS analysis. The size of the carbon blank was found to be linearly correlated with HPLC effluent volume. The blank contribution of the combustion process was indirectly determined by extrapolation of the linear regression line between effluent volume and C blank mass to 0 mL effluent, where it was found to be approximately 1 μg C. Because of their small size ^{14}C measurements were not possible on most of these blanks except for the largest blank sample containing 4.8 μg C, which was found to have a $\Delta^{14}C$ value of -611±26‰ (*8*). This value must be considered a combined purification and combustion blank. In a previous study, CO_2 from several combusted empty tubes was pooled to form two samples large enough for AMS measurements [17]. Average combustion blanks from individual tubes were found to be ~1μg C with $\Delta^{14}C$ of two pooled samples of -623±12‰ and -751±13‰, respectively.

In summary it can be stated that significant amounts of carbon from unknown sources can be added to the samples during preparation. The most crucial steps are the purification procedure and the combustion. The blank associated with the graphitization step of the sample preparation procedure is consistently found to be rather small (~0.1 μg C) (*17, 19, 20*). Some purification procedures, e.g., that used for alkenones, appear to contribute less systematic blank carbon. HPLC preparation may be the strongest source of blank carbon. Little data, however, exist on the other purification procedures. It is important to note that purification, combustion and graphitization blanks are specific for the chemicals, instrumentation, and vacuum-lines used to process the samples, and that they are likely to vary between different laboratories. Thus, it is advisable to process standard material of known mass and isotopic composition to check and if possible correct for blank carbon. AMS results obtained on ultra-microscale samples (<25 μg C) should only be interpreted with great caution. A more conservative approach would only consider data obtained on samples >50 μg C.

3. Interpreting compound-specific radiocarbon contents

When compound-specific or compound-class specific radiocarbon measurements have been obtained and confidence in the results has been established (e.g., blank corrections have been performed), data can be interpreted in various ways. For the purposes of this overview article, we will distinguish between studies where radiocarbon is used as a "tracer" for incorporation of fossil vs. modern carbon, and those where only atmospheric CO_2 or ambient DIC are possible carbon sources, $\Delta^{14}C$ is determined by the decay that has occurred after formation of the compounds and therefore is mostly a measure of the age of the sample.

3.1 Radiocarbon as a tracer - utilization of carbon sources

Several studies have used CSRA in order to investigate the carbon source utilized for the formation of the respective compound. The interest can be from a biological point of view, where the metabolism of a certain organism is investigated. Petsch et al. (*24*) have shown, for example, that prokaryotic organisms assimilate organic carbon from shales during weathering, and that this carbon contributes up to 94% of carbon to the organisms' lipids. For this purpose, the authors isolated phospholipid fatty acids (PLFA), which are considered to be biomarkers for living bacteria, from bacterial cultures grown on rock samples obtained from a weathering profile of a shale outcrop. The $\Delta^{14}C$ levels of these PLFA were significantly ^{14}C-depleted (-711‰ and -922‰), similar to the kerogen value from the shale (-990‰). Similarly, radiocarbon measurements on PLFA isolated from pristine and lignite-polluted soils have been used as evidence for bacterial assimilation of lignite [25]. In the polluted soils, saturated PLFA were significantly depleted in ^{14}C relative to those from the pristine soil.

Radiocarbon can also be a useful tracer if the source of a toxic compound in the environment is unknown. If such a compound derives from industrial pollution, it is likely to be radiocarbon-depleted or radiocarbon-dead as in most cases industrial production is based on fossil carbon sources. In contrast, if the compound originates from a natural source, or if it is produced from combustion of wood (e.g., polycyclic aromatic hydrocarbons; [26]), its radiocarbon levels should be modern. This approach was also taken to elucidate the source of toxic brominated compounds found in high abundance in whale blubber. By isolating these compounds from the blubber of a deceased whale, it was shown that the source of these toxins is most likely not industrial pollution, as the radiocarbon levels were clearly modern [2]. Comparison with industrially produced toxins showed that the latter were strongly depleted in ^{14}C.

3.2 Heterogeneous mixtures

In the introduction it was stated that problems of dealing with heterogeneous mixtures of organic materials when dating TOC samples could be circumvented when analyzing single compounds or compound-classes of known origin. If for instance a biomarker could be isolated from a marine sediment containing large proportions of pre-aged terrigenous organic matter, the ^{14}C-age of this biomarker sample is expected to more accurately reflect the age of deposition of the sample. Indeed we observed a good agreement between radiocarbon ages of planktonic foraminifera, which are commonly used for dating of marine sediment, and those of phytoplankton-derived alkenones in samples from the continental margin of the southern South China Sea [27]. However, radiocarbon ages of planktonic foraminifera and alkenones do not agree quite as well in all locations studied so far. In fact, significant age discrepancies have been observed between the two surface-water derived sediment constituents. The most drastic example has been described by Ohkouchi et al. [28]. The authors studied the Bermuda Rise where age differences between planktonic foraminifera amount to up to 7000 years with the alkenones always being older than the corresponding foraminifera. If based on their larger size and resulting higher settling velocity the latter are assumed to reflect more accurately the age of deposition of the sediment, the offsets can only be explained by admixture of pre-aged allochthonous material. In the study from the Bermuda Rise, carbonate content of the sediment was used as an independent indicator for the contribution of allochthonous material. Age offsets correlate inversely with carbonate content, indicating that the pre-aged organic material is associated with clay-rich allochthonous sediment particles.

The above described results were rather alarming for the community of paleoceanographers who in their search for sites providing archives with high temporal resolution often cored sediment drift bodies. As indicated by the Bermuda Rise data and subsequent similar results [27, 29], re-distribution of pre-aged material is a widespread phenomenon and has often been overlooked.

4. Conclusions

Compound-specific radiocarbon analysis is a powerful tool that helps to elucidate important questions in biogeochemistry, environmental geochemistry, sedimentology, and paleoclimatology. The techniques require application of rather sophisticated purification protocols, and several technical challenges have to be met. These include elimination of any contamination and reduction of blank carbon addition during sample processing to a minimum size. Blank contribution must be carefully evaluated before interpretation of ^{14}C data of small samples. Further systematic research of blanks associated with purification procedures is necessary. Furthermore, dedicated measurement techniques have to be applied involving measurement of size-matched standards in order to correct for isotopic fractionation related to low ion current intensities.

Compound-specific radiocarbon analysis can be used to trace utilization of fossil carbon as a carbon source, and to study provenance of toxins. Applications also include studying radiocarbon ages in soils and sediments in order to understand cycling of carbon, sedimentation processes, and provenance of organic material in the deposits. Important findings have already been made using compound-specific radiocarbon analysis, which has shed new light on processes ignored for a long time.

References

[1] B. E. Rosenheim *et al.*, *Geochemistry, Geophysics, Geosystems, G3* **9**, Q04005 (2008).
[2] E. L. Teuten, L. Xu, C. M. Reddy, *Science* **307**, 917 (2005).
[3] A. Ingalls, R. F. Anderson, A. Pearson, *Marine Chemistry* **92**, 91 (2004).
[4] C. Rabouille, N. Tisnérat, D. Blamart, *Deep Sea Research Part II: Topical Studies in Oceanography* **49**, 1953 (2002).
[5] T. I. Eglinton, L. I. Aluwihare, J. E. Bauer, E. R. M. Druffel, A. P. McNichol, *Analytical Chemistry* **68**, 904 (1996).
[6] A. Pearson, A. P. McNichol, B. C. Benitez-Nelson, J. M. Hayes, T. I. Eglinton, *Geochimica et Cosmochimica Acta* **65**, 3123 (2001).
[7] R. H. Smittenberg *et al.*, *Paleoceanography* **19**, doi:10.1029/2003PA000927 (2004).
[8] S. R. Shah, A. Pearson, *Radiocarbon* **49**, 69 (2007).
[9] E. Calvo, C. Pelejero, G. A. Logan, *Journal of Chromatography A* **989**, 197 (2003).

[10] N. Ohkouchi, T. I. Eglinton, K. Hughen, E. Roosen, L. D. Keigwin, *Radiocarbon* **47**, 425 (2005).
[11] P. J. Müller, G. Kirst, G. Ruhland, I. von Storch, A. Rosell-Melé, *Geochimica et Cosmochimica Acta* **62**, 1757 (1998).
[12] T. I. Eglinton *et al.*, *Science* **277**, 796 (1997).
[13] J. Hwang, E. R. M. Druffel, T. Komada, *Global Biogeochemical Cycles* **19**, GB2018 (2005).
[14] I. T. Marlowe *et al.*, *British Phycological Journal* **19**, 203 (1984).
[15] N. Ohkouchi, L. Xu, C. M. Reddy, D. Montluçon, T. I. Eglinton, *Radiocarbon* **47**, 401 (2005).
[16] R. H. Smittenberg, E. C. Hopmans, S. Schouten, J. S. Sinninghe Damsté, *Journal of Chromatography A* **978**, 129 (2002).
[17] A. Pearson, A. McNichol, R. J. Schneider, K. F. von Reden, Y. Zheng *Radiocarbon* **40**, 61 (1998).
[18] J. S. Vogel, J. Southon, D. E. Nelson, T. A. Brown, *Nuclear Instruments and Methods in Physics Research B* **5**, 289 (1984).
[19] G. M. Santos, J. Southon, S. Griffin, S. R. Beaupre, E. R. M. Druffel, *Nuclear Instruments and Methods in Physics Research B* **259**, 293 (2007).
[20] M. Uchida, Y. Shibata, M. Yoneda, T. Kobayashi, M. Morita, *Nuclear Instruments and Methods in Physics Research Section B: Beam Interactions with Materials and Atoms* **223-224**, 313 (2004).
[21] K. von Reden, A. McNichol, A. Pearson, R. J. Schneider, *Radiocarbon* **40**, 247 (1998).
[22] J. Rethemeyer, University of Kiel (2004).
[23] G. Mollenhauer, D. Montluçon, T. I. Eglinton, *Radiocarbon* **47**, 413 (2005).
[24] S. T. Petsch, T. I. Eglinton, K. J. Edwards, *Science* **292**, 1127 (2001).
[25] J. Rethemeyer *et al.*, *Radiocarbon* **46**, 465 (2004).
[26] A. L. Lima, Dissertation, Massachusetts Institute of Technology, Woods Hole Oceanographic Instiution (2004).
[27] G. Mollenhauer *et al.*, *Paleoceanography* **20**, PA1016; doi:10.1029/2004PA001103 (2005).
[28] N. Ohkouchi, T. I. Eglinton, L. D. Keigwin, J. M. Hayes, *Science* **298**, 1224 (2002).
[29] G. Mollenhauer *et al.*, *Geochimica et Cosmochimica Acta* **67**, 2157 (2003).

Overview of the methods for the measurement and interpretation of short-lived radioisotopes and their limits.

Bassam Ghaleb

GEOTOP-UQAM-McGILL, Université du Québec à Montréal, case postale 8888, succ. Centre-Ville, Montréal, QC, H3C 3P8, Canada

E-mail : ghaleb.bassam@uqam.ca

Abstract. The daughter products of the uranium and thorium series consist of several radioactive isotopes with half-lives varying from less than a second to 10^5 years. Combining their half-live with their geochemical behaviour some of these radioisotopes could be used as tracers and/or chronometers of sedimentary processes. For example, thorium isotopes, and to a lesser extent polonium isotopes are characterized by very low solubility and very high affinity for the surface of particles. Consequently, thorium isotopes can be used to document scavenging and adsorption processes. On the other hand, radium isotopes tend to remain in solution and can be used to document diffusion processes. In the following, we present the analytical methods for the measurement and analysis of the most common short-lived isotopes and throughout their utility in studying sedimentary processes will be illustrated by a few examples of applications. These examples will focus essentially on the applications of short lived thorium isotopes (notably ^{234}Th) and the use of ^{210}Pb as chronometer for recent sedimentary accumulation.

1. Short lived thorium isotopes:

Thorium has three short-lived isotopes ^{234}Th (half-life 24.1 d), ^{227}Th (half-life 18.6 d) and ^{228}Th (half-life 1.91 y). ^{234}Th is a beta emitter while ^{227}Th and ^{228}Th are alpha emitters.

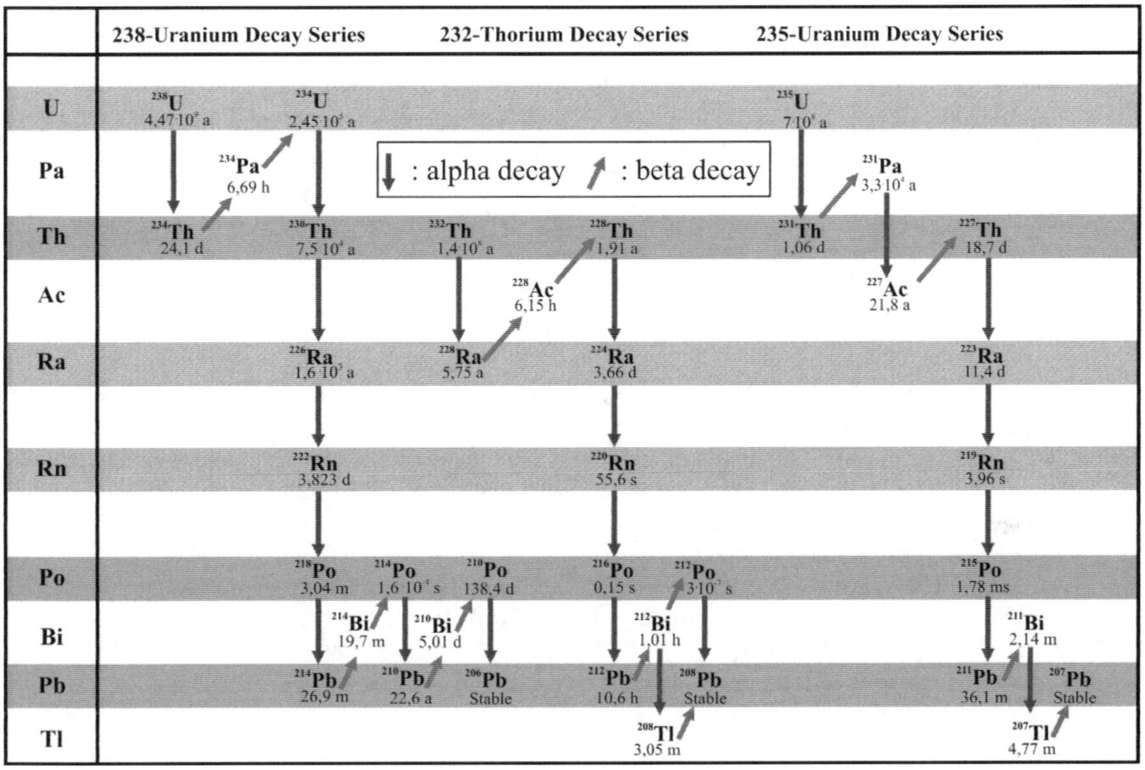

Figure 1. The Uranium and Thorium natural decay series, low solubility radioisotopes are highlighted in pink while those more soluble are highlighted in light blue.

1.1 ^{234}Th: Thorium-234 is produced by the disintegration of ^{238}U. After approximately 120 days in a closed radioactive system, the activity (λ*N) of ^{234}Th will reach a secular equilibrium with that of ^{238}U. This very short half-life makes its determination by mass spectrometric technique very difficult and the counting method remains the method of choice. ^{234}Th is a beta emitter and can be determined using beta detector for which low background is required. The activity in dpm (disintegration per minute) can then be calculated using equation 1.

$$^{234}\text{Th} = \frac{\text{beta counting rate per minute}}{\omega * y} \qquad (1)$$

Where ω is the beta detector efficiency, which is usually determined by the use of a standard for which the activity is known, and y is the chemical extraction yield, which is usually obtained by the use of an artificial tracer isotope such as ^{229}Th.

The other technique widely used for ^{234}Th determination is the gamma emission counter. In this case, ^{234}Th is determined from its gamma emission energy of 63.3 keV. ^{234}Th has a small branching ratio at this energy of approximately 3.8% and this coupled with the relatively low efficiency of the planar gamma detector requires the use of a relatively large sample size. However, the gamma counting is often used for ^{234}Th measurements in seawater samples where it is adsorbed on two MnO_2 cartridges connected in series with several hundred liters of seawater. The development of high purity Germanium well detectors with high efficiency counting allows for the reduction of the sample size.
The activity of ^{234}Th is calculated using equation 2.

IOP Conf. Series: Earth and Environmental Science **5** (2009) 012007 doi:10.1088/1755-1307/5/1/012007

$$^{234}\mathrm{Th} = \frac{\text{gamma counting rate per minute of cartridge (A)}}{\omega * E * V} \qquad (2)$$

Where ω is the gamma detector efficiency determined by the counting of a standard of known activity and by having a matrix similar to that of the samples in order to reduce the self-absorption effect,
V is volume of water pumped through the two (A and B) MnO_2 cartridges,
E is the Mn extraction efficiency and can be calculated from equation 3.

$$E = 1 - \frac{gamma \text{ counting rate in cartridge } Mn(B)}{gamma \text{ counting rate in cartridge } Mn(A)} \qquad (3)$$

The analytical error on the ^{234}Th determinations depends on the number of counts and is generally below 5%.
The applications of ^{234}Th in documenting sedimentary processes are numerous (see Cochran and Masqué, 2003 for a review [1]), but can essentially be summarized in 4 categories:

1.1.1 ***Scavenging processes:***
After the observations of Bhat et al, 1969 [2] that ^{234}Th presents deficiencies relative to ^{238}U in the upper 100 m of the ocean, Matsumoto, 1975 [3] proposed a simple scavenging model based on the balance between the production rate of ^{234}Th from the ^{238}U dissolved in the seawater (assumed to be constant) and the decay combined with the scavenging of ^{234}Th.

$$P = (\lambda + K) \times N \qquad (4)$$

Where P is the atomic production rate of ^{234}Th from ^{238}U dissolved in seawater, λ is the decay constant of ^{234}Th, N is the number of atoms of ^{234}Th and K stands for the scavenging rate constant.
Transforming equation 4 in activity allows for the calculation of K:

$$K = \frac{\lambda \times (P - A_{Th})}{A_{Th}} \qquad (5)$$

From the K value, we can calculate the mean residence time for ^{234}Th (τ_{234Th}) in the water column.

$$\tau_{234Th} = \frac{1}{K} \qquad (6)$$

The results of Matsumoto, 1975 [3] for average deficiencies of ^{234}Th/^{238}U ~ 0.8 in upper 100 m of the Pacific allowed him to calculate τ_{234Th} of ~ 0.4 years. This value is in agreement with the value obtained by Broecker et al, 1973 [4] using ^{228}Th/^{228}Ra disequilibria.

Since the first scavenging model, several more complex models were developed which take into account the dissolved and particulate thorium, adsorption processes, and the size of particles (see Cochran and Masqué, 2003, for a review [1]).

1.1.2 ***Tracing and estimating flux of organic carbon in the oceans:***
Estimating the flux of organic carbon in the oceans is the most common application for ^{234}Th. Coale and Bruland, 1987, [5] (fig.2) showed that there is a direct correlation between the organic pigments and the deficit of ^{234}Th in the euphotic zone of the Pacific. They suggest that higher pigment content corresponds to a lower ^{234}Th/^{238}U activity ratio and that organic matter has a role as an efficient scavenger of ^{234}Th.

IOP Conf. Series: Earth and Environmental Science **5** (2009) 012007 doi:10.1088/1755-1307/5/1/012007

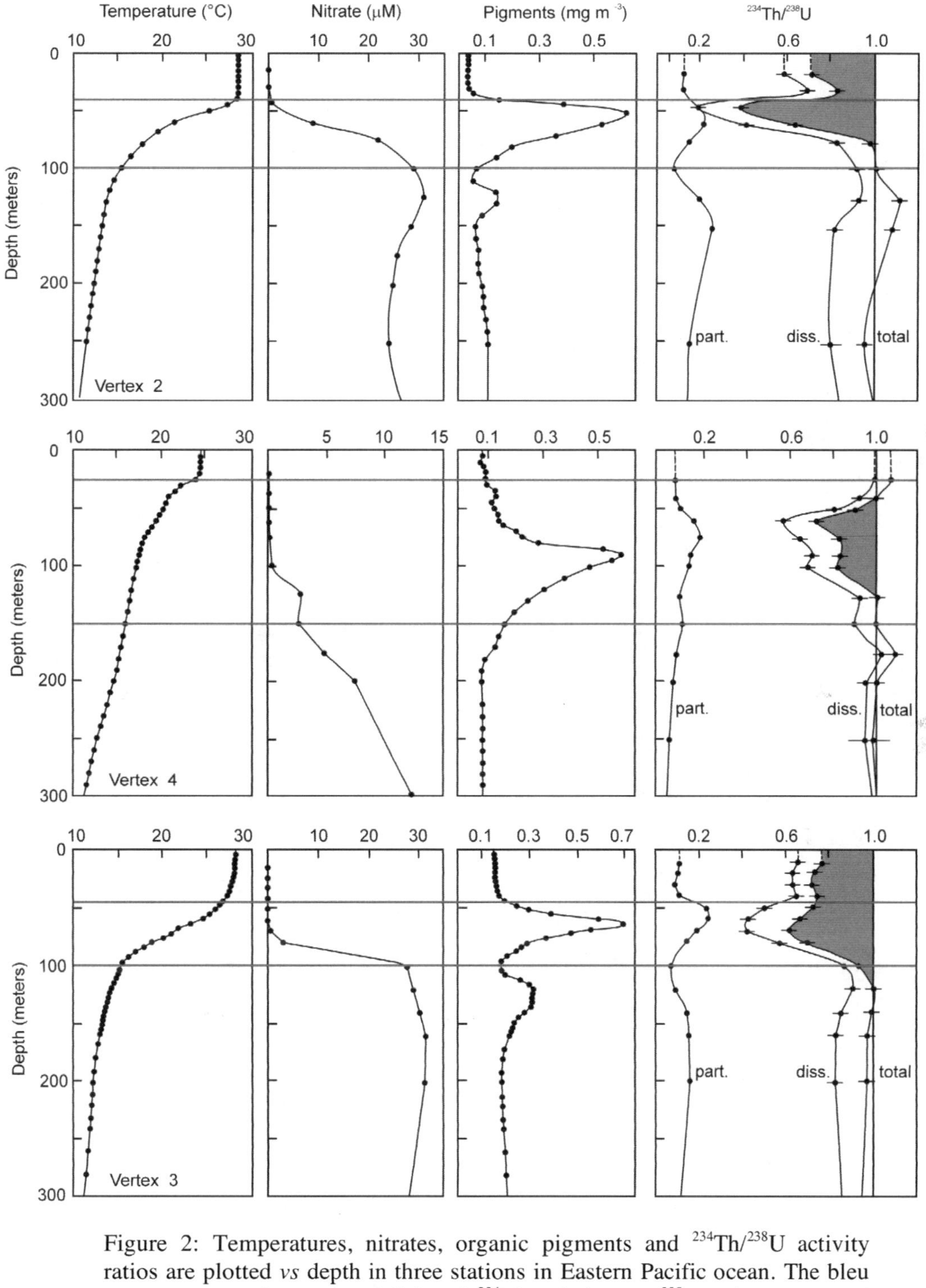

Figure 2: Temperatures, nitrates, organic pigments and $^{234}Th/^{238}U$ activity ratios are plotted *vs* depth in three stations in Eastern Pacific ocean. The bleu area corresponds to the deficit of ^{234}Th relative to ^{238}U. The upper red horizontal line indicates the base of the mixed water mass and the lower red horizontal line represent the limit of the euphotic zone. From Coale and Bruland,1987 [5].

The relationship between the flux of particulate organic carbon (POC) and ^{234}Th is schematized in figure 3. The supply of nutrients in the upwelling zone induces a new production of organic matter by biological bloom. The new organic matter (O.M.) will be exported through sinking particles.

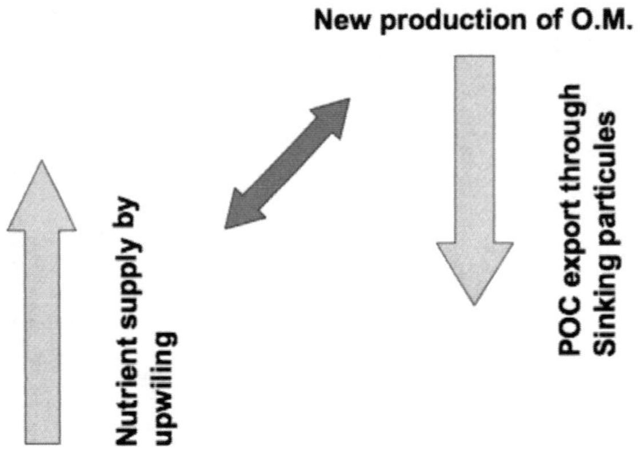

Figure 3: Schema representing the POC flux with sinking particles

Since the work of Coale and Bruland, 1987 [5], several researchers including Ken Busseler (cafethorium.whoi.edu) and his group at Woods Whole Oceanographic Institute focused their efforts on the estimation of carbon flux using ^{234}Th.

Buesseler et al. 1992, [6] proposed a model in which they suggest that the deficit of ^{234}Th with respect to its parent ^{238}U corresponds to the flux of ^{234}Th associated with the sinking particles in the euphotic zone. We can determine the ratio POC/^{234}Th in the sinking particles (from sediment trap for example) then the POC flux can be calculated from the deficit of ^{234}Th using the equation:

$$POC\,flux = \left[\lambda \int (A_U - A_{Th}) \bullet dz\right] \times \frac{POC}{^{234}Th} \qquad (8)$$

Where POC flux is the flux of particulate organic carbon, λ is the decay constant of ^{234}Th, $\int (A_U - A_{Th}) \bullet dz$ is the integrated deficit of the depth of ^{234}Th with respect to ^{238}U and $\frac{POC}{^{234}Th}$ is the ratio of particulate organic carbon to the activity of ^{234}Th measured in the sinking particles.

The JGOFS program (Joint Global Ocean Flux Study) was an excellent opportunity to compare the POC flux derived from ^{234}Th with POC flux from primary production (see Buesseler et al, 1992 [6], for the syntheses of this study)

1.1.3 Indicator of bioturbation

^{234}Th excess disappears in the first centimeter due its short half-life. However, the presence of benthic biological activity at sediment-water interface may allow the penetration of ^{234}Th excess up to the first few centimeters. The mixing and transport processes by living fauna are known as bioturbation and can be estimated from the ^{234}Th profiles , Aller and Cochran, 1976, [7].

1.1.4 Estimation of residence time of suspended particulate mater (SPM)

The affinity of Th isotopes to particles allowed Zhang, 1996, [8] to show the potential of using ^{234}Th to estimate the residence time of suspended particulate matter (SPM) in Saguenay Fjord, Quebec,

Canada. In this environment, the water column is composed of essentially two water masses. The upper water masses (10 to 15 meters) is fresh water which is underlain by thick, salty water. The detrital particles resulting from the erosion of the drainage basin are concentrated at interface between these two water masses and they travel from upstream to downstream with the current of the fresh water. In this context, Zhang, 2000, assumed a constant production rate of ^{234}Th and ^{228}Th from dissolved ^{238}U and ^{228}Ra respectively. In this case, at the head of the fjord prior to any significant ^{234}Th decay (residence time of SPM is ~ 0), a linear relation is observed between the absolute activity of ^{234}Th and ^{228}Th or $^{228}Th_{xs}$ expressed as ^{228}Th-^{232}Th (figure 4).

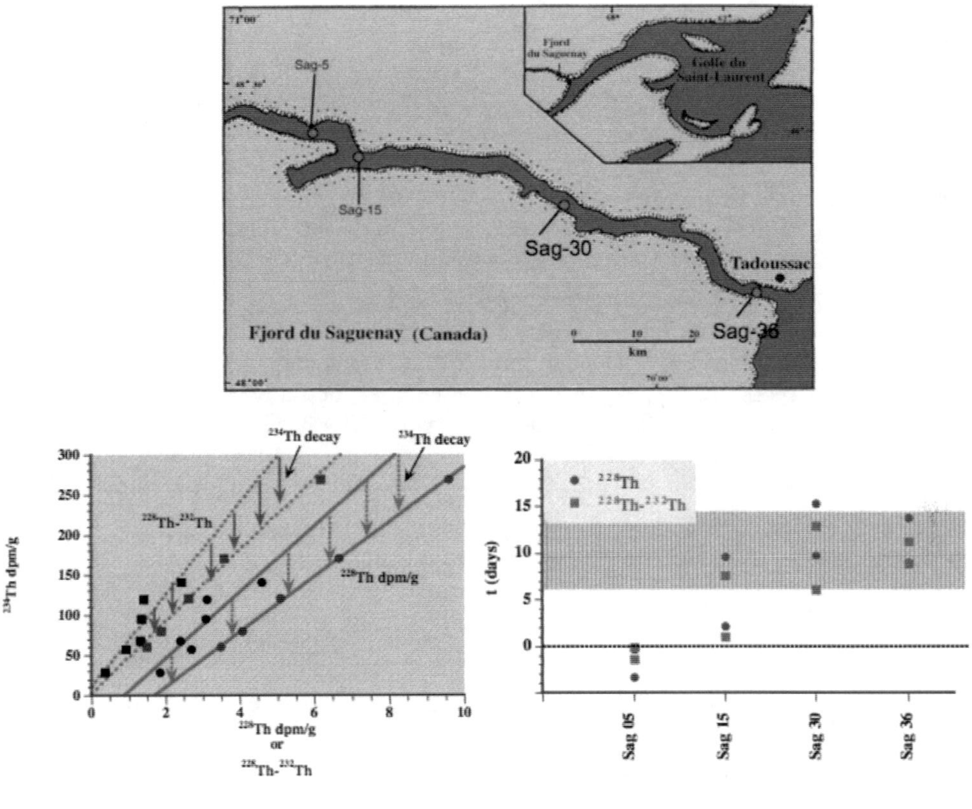

Figure 4: The upper part of the figure show a location map of the analyzed four stations in Saguenay fjord. The lower left side shows the relation between ^{234}Th with ^{228}Th as well with $^{228}Th_{xs}$ expressed as (^{228}Th-^{232}Th). On the right side, the residence time for analyzed stations derived form ^{234}Th model, data from Zhang 2000 [8].

During the transport from upstream to downstream, the SPM will continue to adsorb ^{234}Th and ^{228}Th from dissolved ^{238}U and ^{228}Ra but the ^{234}Th that had been adsorbed earlier at the head of the fjord will show a significant decrease due to radioactive decay (half-life =24.1 d) while the radioactive decay of ^{228}Th is negligible (half-life=1.91 y). Using this, an estimation of the residence time of the SPM can be obtained. The results allow Zhang to estimate the average residence time in the Saguenay fjord from a few days to two weeks, a value in good agreement with that proposed by Laforêt, 1994, [9] who estimated the average residence time of the fresh water mass in the Saguenay fjord to be 6 to 8 days.

1.2 ^{227}Th: ^{227}Th is the direct daughter of ^{231}Pa in the ^{235}U decay chain. It is an alpha emitter at 6.1 Mev and has a half-life of 18.6 days. Its chronological interest is very limited. However, ^{227}Th can be used to calculate the activity of ^{231}Pa assuming secular equilibrium between these two radioisotopes. The latter is an important radio-chronometer (^{231}Pa-^{235}U) for Quaternary chronology. However, because of the technical difficulties for the determination ^{231}Pa its use is limited. Shen, 1996, [10] explained and used ^{227}Th as a chronometer in the couple ^{231}Pa-^{235}U and in the systematic ^{227}Th/^{230}Th for dating cave deposits in China.

1.3 ^{228}Th: ^{228}Th is the direct daughter of ^{228}Ra and the grand daughter of ^{232}Th. The half-life of ^{228}Th is 1.91 y and its alpha emission energy is 5.5 Mev. The chronological interest of ^{228}Th covers a period of 10 to 15 years. An excellent example of the chronological application of ^{228}Th can be found in the work of Koide et al, 1973, [11] in the Santa Barbara basin sediments, where they decomposed the measured ^{228}Th activity into three sources: ^{228}Th unsupported, ^{228}Th growth from ^{232}Th and ^{228}Th growth from ^{228}Ra. Normalizing the value of the activity of each source by the total activity of ^{232}Th, they concluded that the ^{228}Th unsupported/^{232}Th ratio can be used as a chronological tool for decade length time scales.
^{228}Th can also be used to document scavenging processes in a similar manner to ^{234}Th. Kaufman et al, 1981, [12] used ^{234}Th/^{238}U and ^{228}Th/^{228}Ra disequilibria to calculate thorium residence time in seawater. They plotted a Concordia diagram and their results showed a consistent τ_{thorium} from both isotopes. They interpreted the few discordant points as being the result of the mixture of two water masses from different scavenging regimes.
^{228}Th can be useful to indirectly calculate the ^{228}Ra in sediments by analyzing the powder of a sample for ^{228}Th/^{232}Th twice. The first analysis is to be done after the sample collection and the second after a lag time of 2 to 5 years. The initial ^{228}Ra activity can be calculated using equation (9):

$$\frac{^{228}Ra_{activity\ at\ t=0}}{^{232}Th_{activity}} = \frac{\left(\frac{^{228}Th_{activity}}{^{232}Th_{activity}} - 1 + e^{\lambda_{^{228}Th}\bullet t} - \frac{^{228}Th_{activity\ at\ t=0}}{^{232}Th_{activity}} e^{\lambda_{^{228}Th}\bullet t}\right)\left(\lambda_{^{228}Th} - \lambda_{^{228}Ra}\right)}{\lambda_{^{228}Th}\left(e^{\lambda_{^{228}Ra}\bullet t} - e^{\lambda_{^{228}Th}\bullet t}\right)} \quad (9)$$

2 Lead-210 :

^{210}Pb is a natural radioisotope produced by the ^{238}U decay chain with a half-life of 22.3 y. Its half-life makes it suitable to investigate sedimentary processes occurring over a time of about 120 y. This is interesting because most industrial activities and human anthropogenic effects are within the same time scale. The sources of ^{210}Pb activities in natural sediments are made essentially of two components: 1) a supported ^{210}Pb component, produced within the sediments *via* radioactive decay of the ^{238}U chain that produces ^{222}Rn, and 2) an unsupported ^{210}Pb component derived from ^{222}Rn that first diffuses from the earth's surface into the atmosphere and subsequently decays into ^{210}Pb. Unsupported ^{210}Pb is deposited by rainfall and dry fallout. Normally, as the result of radioactive decay, the unsupported ^{210}Pb in a sediment core section presents high values at the surface that decrease with depth (*i.e.* older sediment) Appleby and Oldfield, 1983, [13]. Consequently, it is possible to establish age-depth models in sediment core sections, by quantifying the unsupported ^{210}Pb activity from the surface down to the bottom until it becomes undetectable or negligible Appleby and Oldfield, 1983 [13]. In order to estimate the supported ^{210}Pb fraction from the total ^{210}Pb activity, one measures the ^{226}Ra activity by gamma spectroscopy (by measuring the activity of a short lived daughter product such as ^{214}Pb or ^{210}Bi and assuming a secular radioactive equilibrium between the ^{226}Ra and these daughters). The measured ^{226}Ra activity is assumed to be in secular radioactive equilibrium with supported ^{210}Pb because the radioisotopes between ^{226}Ra and ^{210}Pb have very short half-lives (fig.1). Another option in estimating the supported ^{210}Pb fraction is to estimate it from the bottom of the core sediments, where all the activities are the same within the error ranges. Figure 5 shows an ideal

IOP Conf. Series: Earth and Environmental Science **5** (2009) 012007 doi:10.1088/1755-1307/5/1/012007

theoretical case of core sediments where the activity of ^{210}Pb decreases exponentially with depth from the surface to a constant value of ~ 2 dpmg^{-1}. This value correspond to the value of supported ^{210}Pb at around 75 cm (fig 5.a). Figure 5.b shows the decreasing of the excess of ^{210}Pb (the total activity of ^{210}Pb - the supported value of ^{210}Pb) to a value of 0 dpmg^{-1} while figure 5.c, shows the linear relationship between the natural logarithm of ^{210}Pb excess *vs* depth.

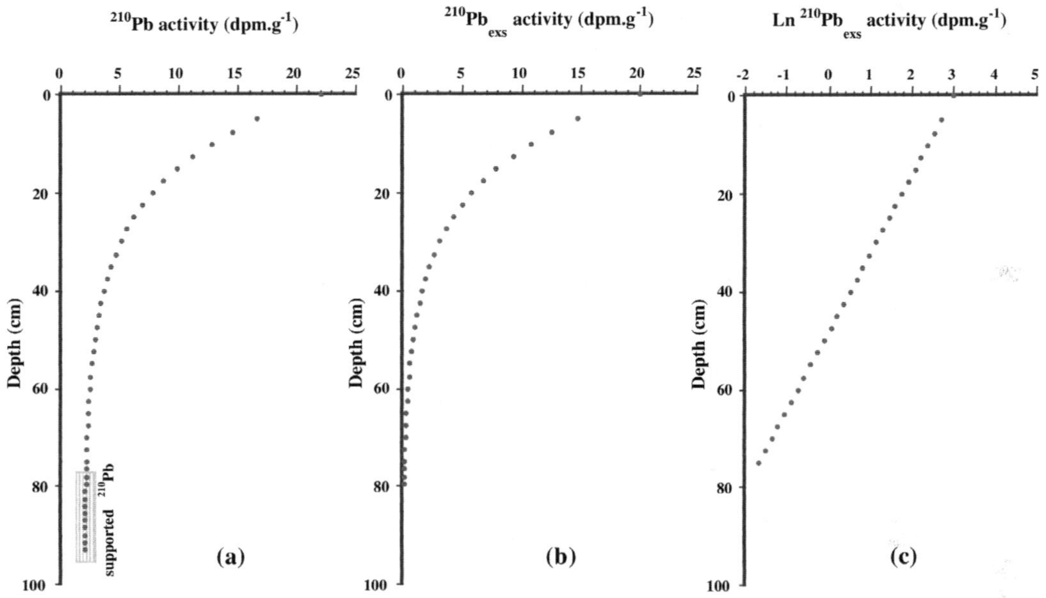

Figure 5: An ideal theoretical ^{210}Pb profile where in (a) the ^{210}Pb activity decreases to the supported ^{210}Pb, in (b) the excess of ^{210}Pb decreases to value=0 and (c) shows the linear relation between the natural logarithm of the ^{210}Pb$_{xs}$ and the time (expressed here as a depth).

2.1 Analytical techniques for ^{210}Pb

2.1.1 Gamma counting emissions:

^{210}Pb can be measured directly for its gamma ray of energy 46.5 keV. However, at this range of low energy, the background is usually relatively high and can represent an important part of the count rate and should be subtracted correctly. Furthermore, the total counting yield is the gamma ray percentage emission (4.1) multiplied by the gamma detector efficiency. Consequently, samples having low radioactivity may require several days of counting to achieve acceptable statistical errors. We should mention that the development of the high purity Germanium well detector (figure 6) improves considerably the gamma detector efficiency hence making the gamma counting emissions method to be used by several more laboratories. It is also worth to mentioning that the gamma counting has the advantage of being a non-destructive method, which enables the sample to be used for other analyses after counting. It also allowed the determination of the total activity of ^{210}Pb (at 46.5 keV) as well as the activity of the supported ^{210}Pb through the determination of the ^{226}Ra daughter product (usually ^{214}Pb and ^{214}Bi). The disadvantage of this method is that gamma ray self-absorption, which depends on the nature of the matrix, requires the sample as a crucial parameter. Therefore, in order to validate the results, we should have some known activity standards that have exactly the same matrix as the analyzed samples.

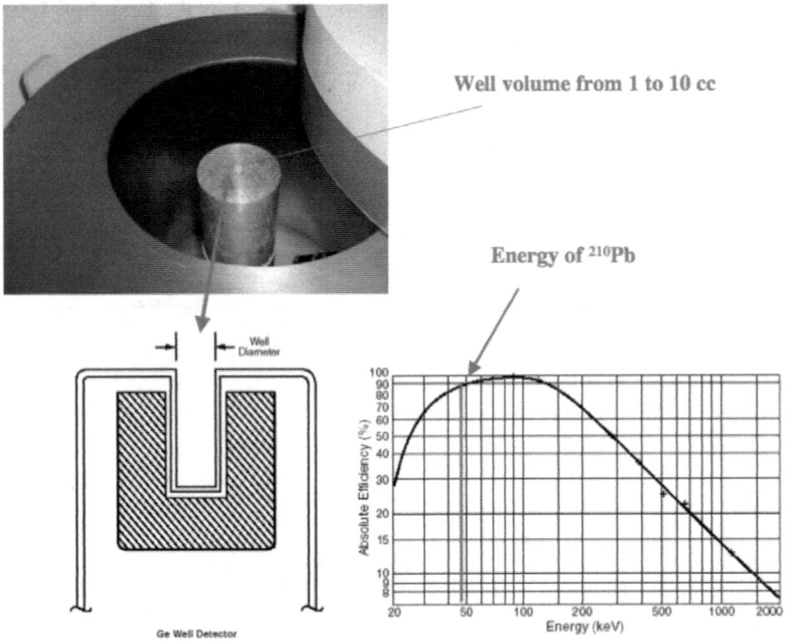

Figure 6: A picture of the gamma well detector at GEOTOP. The lower right side of the figure shows the relation between the absolute efficiency and the gamma energy in the well detector system.

2.1.2 ^{210}Pb determination by α spectrometry *via* ^{210}Po:

This method is based on the assumption that ^{210}Po is in secular equilibrium with ^{210}Pb, which would allow the determination of the activity of ^{210}Po in the samples. The half-life of ^{210}Po is 138.4 days and according to the law of radioactivity, secular equilibrium is reached after ~ 750 days as shown in figure 7, where the activity ratio of ^{210}Po/^{210}Pb evolves with time assuming that at t=0 only ^{210}Pb was present.

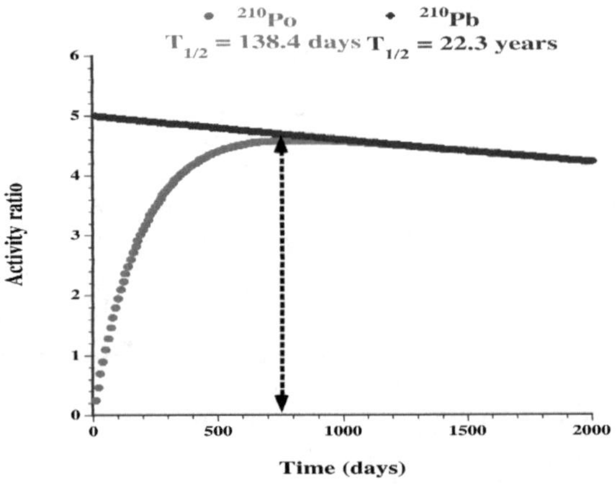

Figure 7: shows the time needed to reach secular equilibrium between ^{210}Pb and ^{210}Po for a sample having only ^{210}Pb at t=0.

IOP Conf. Series: Earth and Environmental Science **5** (2009) 012007 doi:10.1088/1755-1307/5/1/012007

The advantages of this method are: 1) the amount of sample required is small, 2) more sensitive and accurate, 3) the cost of an α spectrometer is cheaper than a γ well detector, and 4) the availability of an artificial radioisotope of polonium (^{209}Po) alpha emitter that can be used as a yield tracer.

2.1.3. Models currently used for interpreting ^{210}Pb depth profiles:

2.1.3.1 CIC model:

Constant initial concentration is the simplest model assuming that the input of the detrital material is constant from year to year, which yields a constant sediment accumulation rate. In this case, the measured activity at depth x, A_x, is related to the initial activity, A_0, according to the equation:

$$A_x = A_0 * e^{-\lambda * t} \qquad (10)$$

Where A_x is the measured activity at depth x, A_0 is the initial activity at depth = 0, λ is the decay constant of ^{210}Pb, and t is the time elapsed since the deposition of sample at depth X. Extracting t from equation 10 yields

$$t = -\frac{1}{\lambda} * \ln\left(\frac{A_x}{A_0}\right) \qquad (11)$$

As mentioned above, when the natural logarithm of $^{210}Pb_{xs}$ is plotted *vs* depth a linear relationship is obtained (fig 5c). The intercept of the line at depth =0 corresponds to the theoretical value of the initial activity of ^{210}Pb at t = 0.
To calculate the sedimentation rate S, we use:

$$S = \frac{X}{t} \qquad (12)$$

Where X is the depth of the cumulated sediment during the period of time t.
By replacing t in equation (12) by its expression in (11) we obtain:

$$S = \frac{X * \lambda}{\ln A_x - \ln A_0} \qquad (13)$$

On the other hand, the slope M of the linear relation between the depth (on X axis) and the natural logarithm of $^{210}Pb_{xs}$(on the Y axis) can be calculated from:

$$M = \frac{\ln A_x - \ln A_0}{X} \qquad (14)$$

By substituting equation (14) in (13) we obtain:

$$S = -\frac{\lambda}{M} \qquad (15)$$

Where S is the sedimentation rate, λ is the decay constant of ^{210}Pb and M is the slope of linear relationship between the depth on X-axis and the natural logarithm of the $^{210}Pb_{xs}$.
Equation 15 can be used to calculate the average sedimentation rate in an environment where sedimentary processes are relatively constant. In this model, we consider that any deviation from exponential decrease is due to sampling after the fact and / or statistical errors and is not the result of a change in sedimentation rate.
It is worth to note that in some cases, we have a sudden change in the net dry mass accumulation rate, such as when a natural lake is transformed into a hydroelectric reservoir. Houel et al, 2005, [14] give an example where usual linear profiles have several segments of non linear profiles derived from lake sediments that has been transformed into hydroelectric reservoirs. They also show that the junction points between the segments indicate the time of the transformation of the lake. Using the slope of each linear segment, they were able to quantify the impact of this transformation in terms of sedimentation rate.

2.1.3.2 CRS model:

The constant rate of supply model was developed in order to take into account the dramatic changes in the last century in the rate of erosion and sedimentation. This model considers that any deviation from a decreasing exponential is due to a change in the sedimentation rate and is not the result of artifacts of sampling or statistical errors. The model assumes a constant supply of $^{210}Pb_{xs}$ and consequently, during high sedimentation and accumulation periods the supply is diluted by this fast sediment accumulation. On the other hand, during periods of low accumulation the value of $^{210}Pb_{xs}$ is higher. By integrating the total inventory of $^{210}Pb_{xs}$, the length of the sediment core and the inventory of each layer, one can calculate the age of each layer according to the equation:

$$t = -\frac{1}{\lambda} * \frac{\sum_0^{\infty} {}^{210}Pb_{xs} * \rho * T - \sum_0^{x} {}^{210}Pb_{xs} * \rho * T}{\sum_0^{\infty} {}^{210}Pb_{xs} * \rho * T} \qquad (16)$$

Where t is the time elapsed since the deposition of layer X, $\sum_0^{\infty} {}^{210}Pb_{xs} * \rho * T$ denotes the total inventory of $^{210}Pb_{xs}$ along the length of the core, $\sum_0^{x} {}^{210}Pb_{xs} * \rho * T$ denotes the total inventory of $^{210}Pb_{xs}$ from the surface to the depth X, ρ is the dry bulk density, T is the thickness of the layer. In figure 7, we present a schematic way to derive the calculation using CRS model.

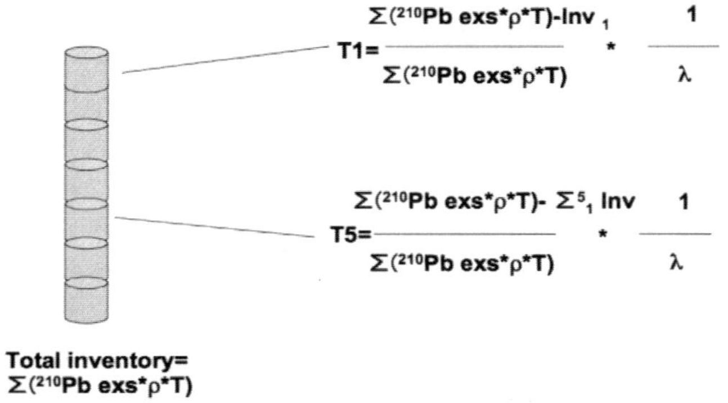

Figure 8: Schematic representation of the CRS model were the age of each layers depend on the inventory of $^{210}Pb_{xs}$, the thickness and the bulk dry density of this layer compared to the total inventory of $^{210}Pb_{xs}$ in the core.

The CRS model was not only applied in lake and marine sediments, but it was widely used in rivers, estuarine and Sabkha environment. It has also been used to estimate peat accumulation rates in peat bogs, Ali *et al* 2008, [15].

2.1.4 Model choice

The choice of the model appropriate for the interpretation of ^{210}Pb results is crucial. As mentioned by Sorgente *et al* 1999, [16] different processes can have similar effects on ^{210}Pb profiles. For example change in the slope of ^{210}Pb profile can be linked to change in sedimentation rate or related to a difference in mineralogy or grain size distribution.

It is obvious that in the case of constant sedimentation rate both models yield the same results. When a significant discrepancy between the two models occurred, the validation of the chronology model usually obtained using independent evidences such as the records of fallout of anthropogenic radioisotopes ^{137}Cs or ^{241}Am.

Conclusion

Short-lived radioisotopes were proved very useful in studies focusing on sedimentary processes with a special attention on recent sediment deposits in the last 150 years. They are also very useful as a tracer and to potentially quantify the flux of particulate organic carbon in aquatic environments. Finally, short-lived isotopes allowed for the understanding and the comprehension of a few fundamental processes such as scavenging, which plays an important role in the distribution and the budget of few chemical elements in the aquatic environment.

With the progress in techniques and the improvement of detection and measurements systems, usually short-lived isotopes as well as new natural and artificial radioisotopes will play an important role in a better understanding and comprehension of aquatic systems.

IOP Conf. Series: Earth and Environmental Science **5** (2009) 012007 doi:10.1088/1755-1307/5/1/012007

Acknowledgements: I'm grateful to Drs Jean David (GEOTOP-Montréal) and Hélène Isnard (CEA-France) for their constructive comments, Caroline Peckler and Christelle Not (GEOTOP-Montréal) for their help.

References

[1] Cochran J K and Masqué P 2003 Short-lived U/Th series radionuclides in the ocean: Ttracers for scavenging rates, export fluxes and particles dynamics *Uranium-series geochemistry (Reviews in Mineralogy & Geochemistry* vol 52) eds B Bourdon, G M Henderson, C C Lundstrom and S P Turner (Washington, U.S.A) chapter 11 pp 461-492

[2] Bhat S G, Krishnaswami S, Lal D, Rama and More W S 1969 *Earth. Planet. Sci. Lett.* **5** 483-491

[3] Matsumoto E 1975 *Gechim. Cosmochim. Acta.* **39** 205-212

[4] Broecker W S, Kaufman A and Trier R M 1973 *Earth. Planet. Sci. Lett.* **20** 35-44

[5] Coale K H and Bruland K W 1987 *Limnol. Oceanogr.* **32** 189-200

[6] Buesseler K O, Bacon M P, Cochran J K and Livingston H D 1992 *Deep-Sea Res.* I **39** 1115-1137

[7] Aller R C and Cochran J K 1976 *Earth. Planet. Sci. Lett.* **29** 37-50

[8] Zhang D 2000 *Flux de radio-isotopes à courte périodes dans les bassin marins marginaux de l'est canadien* thèse de doctorat 193 p, Unversité du Québec à Montréal, Montréal, QC Canada.

[9] Lafôret R G 1994 *Oceanographic and acoustic implications of fresh water release into the Laurentian Channel* M.Sc thesis 95 p, Royal Rooads Military College Victoria, BC, Canada.

[10] Shen G 1996 *Quat Sci. Rev.* **15** 699-707

[11] Koide M, Bruland KW and Goldberg E D 1973 *Gechim. Cosmochim. Acta.* **37** 1171-1187

[12] Kaufman A, Li Y-H and Turekian K K 1981 *Earth. Planet. Sci. Lett.* **54** 385-392

[13] Appleby P G and Oldfield F 1983 *Hydrobilogia* **103** 29-35

[14] Houel S, Louchouarn P, Lucotte M, Canuel R and Ghaleb B 2006 *Limnol. Oceanogr.* **51** (3) 1497-1513

[15] Ali A A, Ghaleb B, Garneau M, Asnong H and Loisel J 2008 Appl. Radiat. Isotopes 66 1350 1358.

[16] Sorgente D, Frignani M, Langone L and Ravaioli M 1999 *Chronology of marine sediments-interpretation of activity depth profiles of ^{210}Pb and other radioactive tracers* Technical report n. 54, 32 p, Instituto per la geologia marina, Bologna.

IOP Conf. Series: Earth and Environmental Science **5** (2009) 012008 doi:10.1088/1755-1307/5/1/012008

The U-series dating of (biogenic) carbonates

C Hillaire-Marcel
GEOTOP, Université du Québec à Montréal, Montréal (Qc) H3C 3P8, Canada

E-mail: geotop@uqam.ca

Abstract - U- and Th-series offer a large array of isotopes that constitute magnifying "lenses" into recent temporal dimensions of Earth System processes. Whereas mass spectrometry (MS) measurements of ^{238}U-^{234}U-^{230}Th and ^{235}U-^{231}Pa disequilibria give access to time ranges varying between about a million of years to hundreds of thousand years, MS or counting methods of shorter-lived daughter isotopes (e.g., ^{226}Ra, ^{210}Pb, ^{234}Th, ^{228}Th, ^{228}Ra) inform on time scales varying from 50 ka (^{226}Ra-^{230}Th "pseudo-concordias"), 10 ka (^{226}Ra-excess method), 100 a (^{210}Pb-^{226}Ra "pseudo-Concordia" or ^{210}Pb-excess method), 30 a (^{228}Ra-excess), 10 a (^{228}Th-^{228}Ra-^{232}Th disequilibria) and up to 3 months (^{234}Th-excess). U-series isotopes, and especially the sequence ^{238}U-^{234}U-^{230}Th revealed essential in validating the astronomical theory of climate through the dating of high interglacial sea levels and provided the means to calibrate the radiocarbon time scale into "calendar years". Shorter-lived natural isotopes of these series (*e.g.*, ^{210}Pb, ^{226}Ra) also revealed essential for the documenting of more recent processes. Here, we examine U-series systematics in biogenic minerals with special attention to biogenic carbonates. We also discuss the impact of diagenetic conditions on dating applications.

1. Introduction

After nearly 60 years of development, since the early work of the Russian school of geochemistry, with thousands of publications, benchmark monographies (e.g. [1]), one could think that all has been said about the "U-series dating method" of chemical or biogenic minerals and compounds. This is not the case. Firstly, because progresses in analytical methods kept opening new applications. From this view point, one should not any more refer to "the U-series dating method", but to " U- (and Th-) series dating methods", since more precise measurement of U and Th daughter isotopes provide now access to a large array of radiometric clocks. Secondly, when two of these radiometric methods overlap, they offer the means to document the relative closure of radioactive systems through time. Thirdly, a better knowledge of the geochemical behavior of the most "mobile" elements of the U and Th decay series (Table 1), notably of U and Ra, now provides stronger grounds to assess the geological significance of U- & Th-series ages, and often leads to model-age dating in situations where this chemical closure has not been maintained.

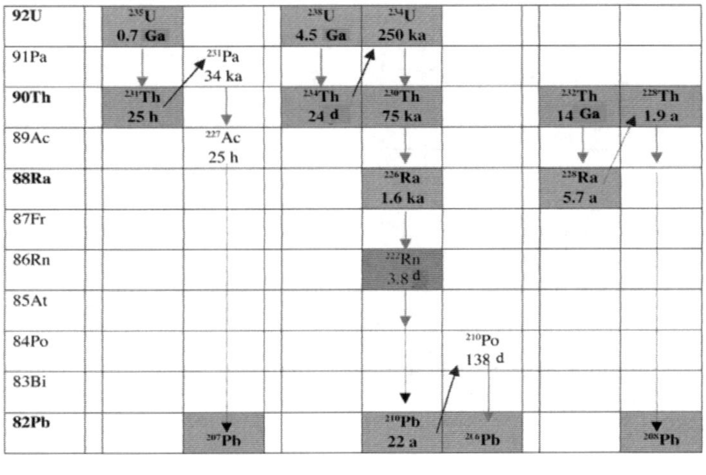

Figure 1: Elements and isotopes in the U & Th decay series of current radiometric use. Blue background: elements showing some solubility, thus with relatively high residence time in natural waters; note that U is a redox sensitive element leading to its precipitation under low-O2 and reduced conditions. In brown: particle-reactive elements with a short residence time in natural waters. In red: the only gaseous element (radon). Numbers in Ga, Ma, ka, a, d and h represent first

order values of the corresponding isotope half-life (in gigayears, megayears, kiloyears, years, days and hours, respectively).

In this chapter, it is not intended to review all U- & Th-series tools, neither to examine all geological and environmental applications of such tools. Recent review articles have been published [2] and the reader is invited to complement his/her readings, if needed, with such reviews. By focusing on the dating of biogenic carbonates, it is intended to examine the various radiometric clocks that can be used and/or combined, and to assess to which extend the mobility of radionuclides in such material may hamper or not their precise dating, thus the setting of ages for host geological units. In a similar fashion, the reader will find, elsewhere in this volume [3] information about some current analytical methods for the measurement of abundances or activities of U- and Th-series radionuclides.

2. The U-series and their potential use as radiochronometers

The most widely used radiochronometers are those of the ^{238}U-series, sometimes combined with information from the ^{235}U-series, more rarely, from the ^{232}Th-series (Fig. 2). However, all isotopes illustrated in Figure 1 are of use in Earth sciences and/or environmental studies, including the shortest-lived of them that may provide key information on fast reaction rate processes (e.g., primary productivity, carbon fluxes...). In the present chapter, more attention is paid to the control of time over longer time scales, those mostly used to document Earth system processes that involves forcing mechanisms operating at 10 to 10^6 years time scales, e.g., from solar to astronomical time scales.

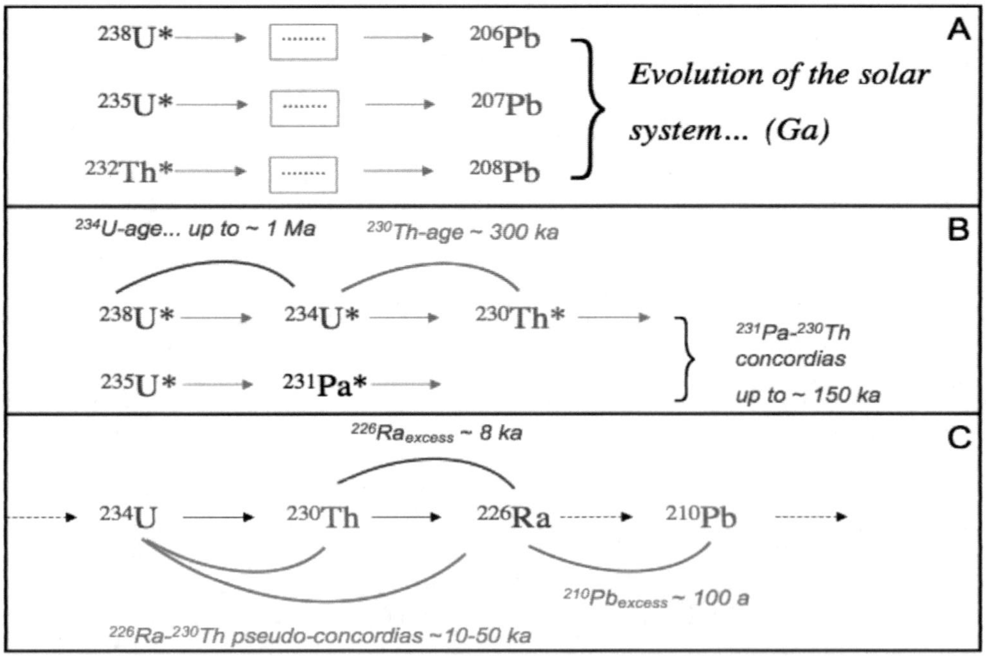

Figure 2: Time scales controlled by various sets of isotopes in the U- and Th-series; color codes are similar to those of Figure 1. A: The U-Pb geochronology that has been used for the dating of very old rocks; B & C: Disequilibria in the U-series providing time control tools over the last million years.

As illustrated in Figure 2A, above, a most important use of these radiochronometers has been the dating of ancient rocks through uranium-lead geochronology [4] using U-rich resistant minerals, such as zircons, that are more likely than other minerals to have maintained a closed geochemical system through time, thus fulfilling a prerequisite for radiometric dating: the chemical closure of the radioactive system since mineral formation. In such "closed" systems, losses and gains are only due to radioactive decay and subsequent growth of daughter nuclides. However, such ideal conditions are rarely preserved in low temperature minerals, such as biogenic carbonate. Firstly, these minerals often show instability through time, with aragonite converting into calcite, more stable at low temperatures, and/or large-scale crystal reorganization in calcite. Secondly, through diagenetic processes, chemical U-uptake or losses may occur, thus upsetting the radiometric clock. Nonetheless, the U- and Th-series provide a large array of radiometric clocks, illustrated in Figures 2B and 2C, which revealed suitable to investigate Earth surface processes that occurred up to nearly two million years, in rare settings [5], thus over shorter periods of time when chemical closure may have prevailed. When it did not, models accounting for chemical mobility of given elements or isotopes may also be proposed (e.g. [6]). These dating tools are often referred to as "U-series disequilibrium methods", although, the Th-series also offer similar clocks, but essentially through the ^{232}Th-^{228}Ra-^{228}Th sequence spanning the last 30 years or about, which are mostly used in chemical oceanography and for recent sedimentation rate studies.

The first key to understand how these U- and Th-series chronometers work for dating the recent Earth history is to refer to the chemical behavior, and particularly the relative solubility of elements from the decay series. In its VI+ state, under oxidizing conditions, U is fairly soluble in most natural waters, about 10^6 times more than Th for example, thus showing a high residence time in the ocean (about >300 ka; [7]). As a consequence, carbonates and biogenic minerals formed in most natural water systems are practically deprived of Th, but that produced through the in situ decay of U through time. In opposition, Th, Pb and Pa are strongly particle-reactive and show very low residence times in natural waters (see table below for the ocean). Therefore, all minerals formed in natural waters may incorporate soluble elements of the U-series in their lattice, such as U or Ra, thus setting all related clocks (Fig. 2), back to 0. A noticeable exception here may be found in highly mineralized and hyperalkaline hydrothermal systems as well as in phosphate-rich deposits, where Th may be present in some abundance in minerals formed under such conditions.

Atomic Number	Symbol	Average molar mass (g)	Conc. (seawater) in nmole.kg^{-1}	Residence Time (a)
82	Pb	207.2000	1.3 10^{-2}	80
84	Po		3.0 10^{-11}	100
86	Rn		3.0 10^{-12}	0.01
88	Ra	226.0254	5.75 10^{-7}	-----
90	Th	232.0381	8.62 10^{-5}	45
91	Pa	231.0359	1.8 10^{-8}	200
92	U	238.0290	1.34 10^{1}	400 10^{3}

From Nozaki, Y., http://www.agu.org/eos_elec/97025e.html

Table 1: Elements of the U and Th series in the ocean

The second key, to understand how U-series disequilibrium dating works concerns U-sensitivity to redox conditions and the physical fractionation of its isotopes when leached/dissolved from minerals, at rock/water interface. In the ^{238}U early decay products, ^{234}U results from ^{234}Th-decay, itself ensuing from the highly energetic decay of ^{238}U with the loss of one ^{4}He nucleus. Such nuclear reactions result in damages of the site occupied by the nuclide, thus to a higher vulnerability to leaching and dissolution of daughter products (e.g. [1, 2]). As a consequence, a relative enrichment in ^{234}U vs. ^{238}U is generally observed in pore waters. A second important property of U here is the low solubility of U under its IV+ state, i.e., under low redox conditions. Thus, when crossing redox boundaries, U precipitates (as Fe and Mn do). It shows typical redox-sensitive distributions in shallow sediments from aquatic environments (e.g., [8]). One consequence of both processes is that deep reduced ground waters depict very low U-concentrations, but up to a nearly 20-fold excess in ^{234}U, vs. its activity when in secular equilibrium with ^{238}U (e.g., [9]). As a consequence, continental waters generally show relative excesses in ^{234}U and through river fluxes of ^{234}U-enriched uranium, account for a substantial part of the near 15% excess in ^{234}U (vs. its concentration when in secular equilibrium with its parent ^{238}U) observed in the ocean [10]. Thus, added to other decay-series chronometers, disequilibria in the ^{238}U-^{234}U pair of isotopes may provide some time control over the 10^4-10^6 a time range (Fig. 2).

Isotope	Atomic mass	Half-life
^{238}U	238.05078	$4.468\ 10^9$ a
^{236}U	236.04558	$2.34\ 10^7$ a
^{235}U	235.04392	$7.13\ 10^8$ a
^{234}U	234.04095	$2.45\ 10^5$ a
^{233}U	233.03963	$1.59\ 10^5$ a
^{232}U	232.03713	68.9 a
^{231}Pa	231.03588	$3.43 * 10^4$ a
^{234}Th	234.04359	24.1 d
^{232}Th	232.03805	$1.39\ 10^{10}$ a
^{230}Th	230.03313	$7.52\ 10^4$ a
^{229}Th	229.03176	$7.3\ 10^3$ a
^{228}Th	228.02872	1.913 a
^{228}Ra	228.03106	5.76 a
^{226}Ra	226.02540	$1.602\ 10^3$a
^{222}Rn	222.01757	3.825 d
^{210}Po	209.98285	138.4 d
^{210}Pb	209.98416	22.3 a

Table 2. Atomic mass and half-lives of the most important isotopes of the U- and Th-decay series from a compilation of the most recent values reported in literature.

A first order estimate of the time range controlled by any given set of radionuclides, with a relative analytical precision better than 3% when determining their concentrations, is given by 5 times the half-life of the shorter-lived isotope involved, as will be seen below.

The analytical methods of current use for such isotopic measurements range from emission spectrometry (α, β, γ), to thermal ionization mass spectrometry (TIMS) and inductively coupled plasma mass spectrometry (ICP-MS). These methods yield analytical precisions from a few per cent, to a few per mil in general.

Half lives of the most important isotopes in the U- and Th-decay series are listed in Table 2 and time ranges covered by the most frequently used methods are illustrated in Figure 3.

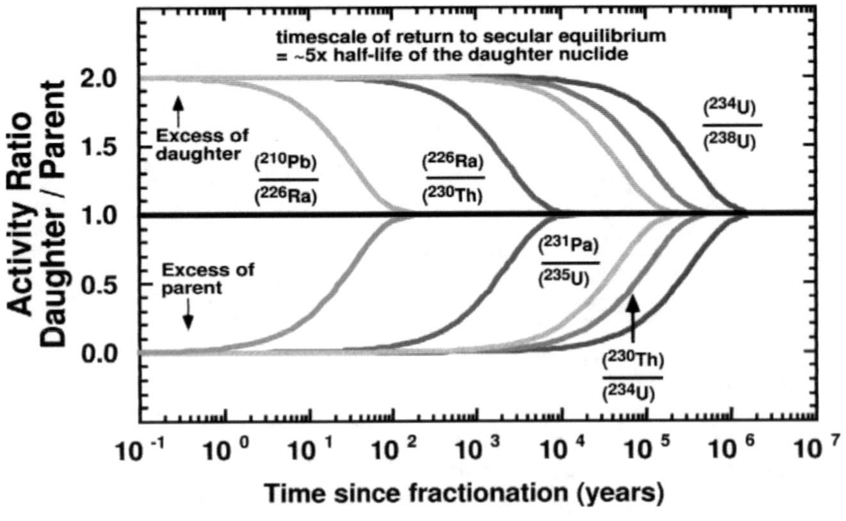

Figure 3: Time ranges covered by the most used disequilibria in the U-Th series methods (from [2]). See text for more explanation on excess vs. deficit in a parent or daughter isotope, and for the definition of activity ratios. The time range (~ 10-50 ka) that can be controlled with the $^{226}Ra/U$ vs $^{230}Th/U$ pseudo-concordia is not illustrated (see text).

3. Basic equations governing radioactive decay

All chapters and monographies addressing radiometric methods present more or less simplified versions of the equations governing activity ratios or relative concentrations of isotopes/elements in decay series. They are summarized below. Excel working sheets for the calculation of ages, based on these equations, are available on the GEOTOP Web site (www.geotop.ca).

For a radioactive element/isotope N:

$$dN = - \lambda_n N * dt$$

where N is the concentration/abundance of the element in a given chemically closed system, "t" represents time and λ_n the decay constant of the element N (in t^{-1} units); ($\lambda_n N$) represents the "activity" of the element in units defined by those used for N and t.

When integrating: $N = N_0\, e^{-\lambda_n t}$ or $\lambda_n N = \lambda_n N_0\, e^{-\lambda_n t}$, where N_0 is the initial concentration of the element. The half-life of the element (T) is defined for $N = N_0/2$, i.e., when $T = \mathrm{Ln}\,(2)/\lambda_n$
When the daughter isotope/element is also radioactive:

$$dN_2/dt = \lambda_1 N_1 - \lambda_2 N_2$$

by integrating, it comes that:

$$N_2 = (N_2)_0 e^{-\lambda_2 t} + \frac{\lambda_1 (N_1)_0}{\lambda_2 - \lambda_1}(e^{-\lambda_1 t} - e^{-\lambda_2 t})$$

In systems where $(N_2)_0 = 0$ (no daughter product at the origin):

$$\frac{\lambda_2 N_2}{\lambda_1 N_1} = (\frac{\lambda_2}{\lambda_2 - \lambda_1}) * (1 - e^{-(\lambda_2 - \lambda_1)t})$$

Applied to the U-series isotopes presented above, this equations becomes:

$$[1]\ \frac{\lambda_{231}{}^{231}Pa}{\lambda_{235}{}^{231}U} = (\frac{\lambda_{231}}{\lambda_{231} - \lambda_{235}})[1 - (e^{-(\lambda_{231} - \lambda_{235})t})]$$

$$[2]\ \frac{\lambda_{234}{}^{234}U}{\lambda_{238}{}^{238}U} = (\frac{\lambda_{234}}{\lambda_{234} - \lambda_{238}}) * (1 - e^{-(\lambda_{234} - \lambda_{238})t})$$

In this later case, one frequently refers to a "$\delta^{234}U$" value defined as:

$$[3]\ \delta_0{}^{234}U = [(\frac{\lambda_{234}{}^{234}U}{\lambda_{238}{}^{238}U})_0 - 1] * 10^3$$

Here, $(\lambda_{238}{}^{238}U)_t \approx (\lambda_{238}{}^{238}U)_0$, due to its slow decay rate in comparison with that of ^{234}U, and since $\lambda_{234} >> \lambda_{238}$, it comes that:

$$[4]\ \delta_0{}^{234}U \approx \left([(\frac{\lambda_{234}{}^{234}U}{\lambda_{238}{}^{238}U})_t - 1] * e^{(\lambda_{234} * t)} \right) * 10^3$$

This equation illustrates the fact the time control is essentially governed by the decay constant of the daughter product when it is much higher than that of the parent isotope

$$\frac{\lambda_2 N_2}{\lambda_1 N_1} \approx (1 - e^{-\lambda_2 t}) \text{ when } \lambda_2 >> \lambda_1$$

When three isotopes or elements are related:

$$\frac{\lambda_3 N_3}{\lambda_2 N_2} = (\frac{\lambda_1 N_1}{\lambda_2 N_2})(1 - e^{-\lambda_3 t}) + (\frac{\lambda_2 N_2 - \lambda_1 N_1}{\lambda_2 N_2})(\frac{\lambda_3}{\lambda_3 - \lambda_2})(1 - e^{-(\lambda_3 - \lambda_2)t})$$

as it is the case in the "historical" U-Th disequilibrium dating method:

$$[5]\ \frac{\lambda_{230}{}^{230}Th}{\lambda_{234}{}^{234}U} = (\frac{\lambda_{238}{}^{238}U}{\lambda_{234}{}^{234}U})(1 - e^{-\lambda_{230} t}) + (\frac{\lambda_{234}{}^{234}U - \lambda_{238}{}^{238}U}{\lambda_{234}{}^{234}U})(\frac{\lambda_{230}}{\lambda_{230} - \lambda_{234}})(1 - e^{-(\lambda_{230} - \lambda_{234})t})$$

Through iterative calculation, the value of "t" can be obtained from this equation.

4. Analytical methods

Techniques for the analysis of U-series isotopes are either based on the measurement of their abundance (mass spectrometry) or of their "activity" (emission spectrometry). In this later case, depending the mode of decay of the element or isotope concerned, one measures the α, β and/or γ emission linked to the nuclear reaction giving birth to the daughter product. Such techniques are illustrated in the method sections of all papers (see also [3]). In most cases, authors using emission spectrometry report activity values with $\pm 1\sigma$ uncertainties based on counting statistics.

A much better precision using smaller quantities of samples can be achieved with direct mass spectrometric measurements either (TIMS or ICP-MS). Here, most authors report isotopic abundances with a $\pm 2\sigma$ uncertainty representing more or less elaborated estimates of the analytical reproducibility.

Again, method sections in published papers should provide the information needed to assess the true significance of such estimates of the analytical uncertainty. Never, this value can be seen as providing directly, through the equations above, the uncertainty on the age of the geological formation sampled. The scatter of ages of the representative samples depends upon a large array of variables. A robust paper addressing this issue is that of Ken Ludwig, in the volume edited by Bourdon et al. [2].

5. Open vs. closed systems

In the equations [1] to [5] above, one prerequisite is that the radioactive system is chemically closed: no gain or loss in any isotope but those linked to the in situ radioactive decay. This is unfortunately rarely the case of biogenic carbonates, where some chemical U-mobility often prevails, particularly during early diagenetic stages. Nonetheless, a few criteria may be used to assess the (relative) closure of the radioactive system. The most unquestionable are based on the coupling of two radioactive clocks spanning at least partly the same time range.

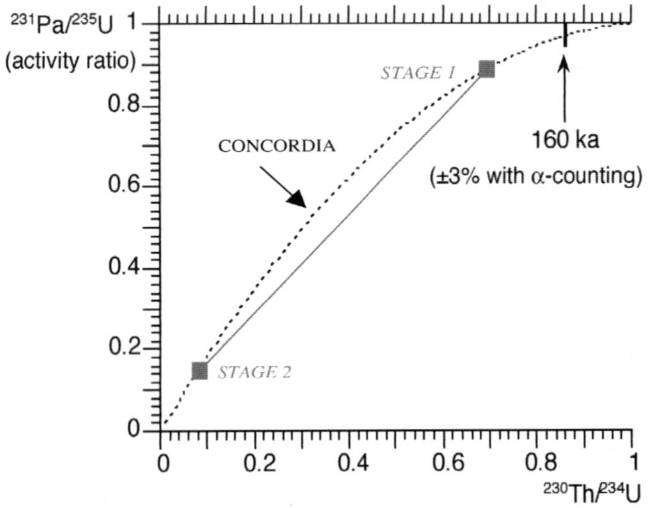

Figure 4: The $^{231}Pa/^{235}U$ vs. $^{230}Th/^{234}U$ concordia. In a closed system, the sample will move, through time, along the "concordia" curve. In red, a theoretical mixing line when two U-uptake stages occur (e.g., Stage 1 = Last Interglacial & Stage 2 = Holocene).

Sometimes, the initial concentrations of the parent elements or isotopes are well known (e.g., ^{238}U vs. ^{235}U in natural uranium, linked through their mass ratio of 137.88 in recent rocks). In such cases, the temporal evolution of $^{230}Th/^{234}U$ (^{238}U) vs. $^{231}Pa/^{235}U$ ratios define a "Concordia" curve. All samples that remained chemically closed with respect to U should fit on this curve. When, for example, two stages of U-uptake occurred, samples representing various proportions of these two components should follow a linear relationship whose end-members on the curve represent the ages of the two stages (Fig. 4 above).

Pseudo-concordias can also be used by combining information from several isotopes from the ^{238}U decay-series. For example, when a given $^{234}U/^{238}U$ ratio can be postulated for the uranium initially present in the sample, as illustrated in the abaca of Figure 5, the sample should evolve through time along a given curve.

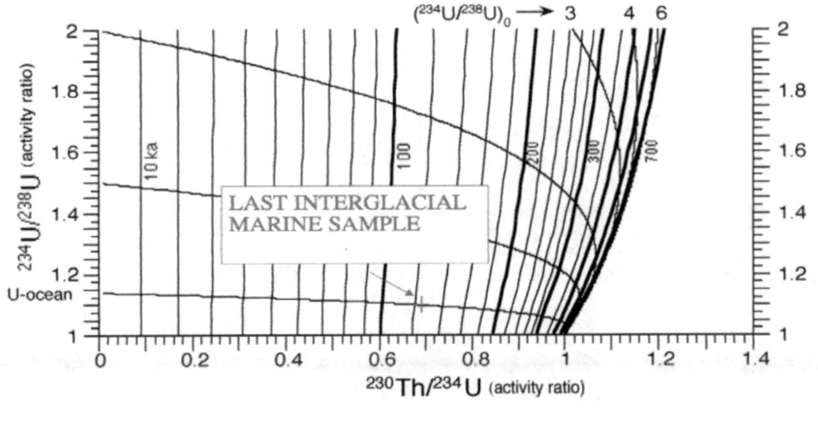

Figure 5: Abaca illustrating the evolution of activity ratios in the ^{238}U series, through time, for a given initial $^{234}U/^{238}U$ activity ratio. This is the case of uranium in the ocean that shows a quite constant $^{234}U/^{238}U$ activity ratio (~1.15). Thus, a marine sample from the Last Interglacial should fit on the "pseudo-concordia"

curve defined by this initial U-composition, as illustrated here.

In a relatively narrow temporal domain, ^{238}U-^{234}U-^{230}Th and ^{226}Ra, can also be used to document the closure of the radioactive system as illustrated in Figure 6 below. The ingrowth of ^{230}Th and of its daughter ^{226}Ra in a hypothetical mineral that would contain only U when formed, should follow the curves of Figure 6A, thus allowing to control any U-mobility during a few tens of years. However, most minerals contain some radium (whose geochemistry follows closely that of barium). This initial radium in excess (i.e., unsupported by parent ^{230}Th, as illustrated in Fig. 6B) decays rapidly, due to the high decay constant of ^{226}Ra, and ^{226}Ra returns to equilibrium with ^{230}Th within less than 10 ka. Later on, the pseudo-concordia can still be used in the 10-50 ka time range. Any diagenetic addition or subtraction of U (and/or of Ra) will move the datum point above or below the curve of Figure 6B.

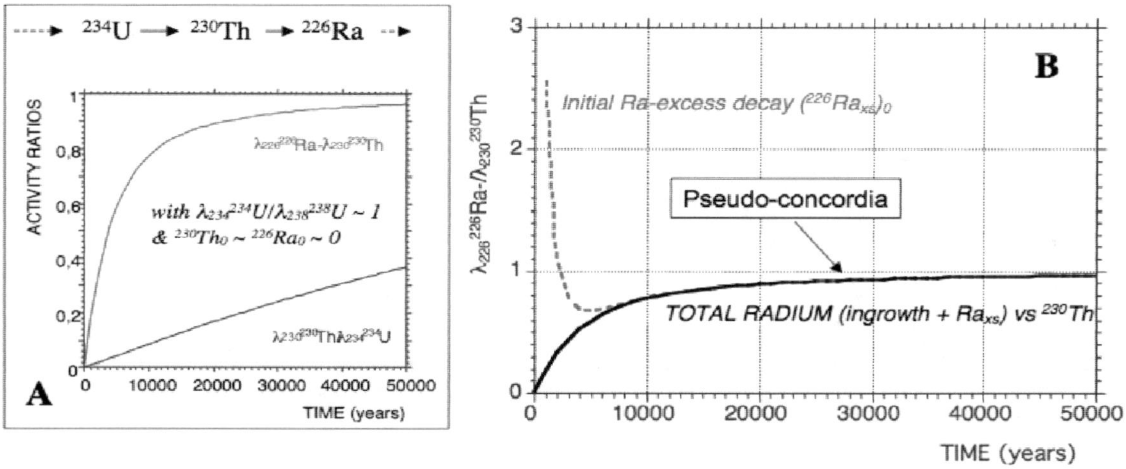

Figure 6: ^{226}Ra vs. ^{230}Th "pseudo-concordia". A: Growth curves of ^{230}Th and ^{226}Ra from parent U in a closed system; B: Decay of the initial excess in ^{226}Ra and pseudo-concordia $^{226}Ra/^{230}Th$.

6. The "detrital" contamination of samples

Another limitation when using equations [1] and [5] above concerns the assumption that the initial contents in ^{231}Pa and ^{230}Th daughter products, respectively, was null, i.e., that their concentration after time "t" is solely due to radioactive ingrowth. However, it is quite common to observe samples "contaminated" by detrital fractions containing variable amounts of U and Th series isotopes, with variable disequilibria linked to the genetic and transport history of the contaminating minerals and particles. In such a condition, equations [1] and [5] should be modified in order to include a term representing the "initial content" in daughter products, as follows:

$$[6]\ \lambda_{231}{}^{231}Pa = \left[\lambda_{231}({}^{231}Pa)_0(e^{-\lambda_{231}t})\right] + \left(\lambda_{235}{}^{235}U\right)\left[\left(\frac{\lambda_{231}}{\lambda_{231}-\lambda_{235}}\right)[1-(e^{-(\lambda_{231}-\lambda_{235})t})]\right]$$

$$[7]\ \lambda_{230}{}^{230}Th = \left[\left(\lambda_{230}\left({}^{230}Th\right)_0\right)\left(e^{-\lambda_{230}t}\right)\right]$$
$$+\left(\lambda_{234}{}^{234}U\right)\left[\left(\frac{\lambda_{238}{}^{238}U}{\lambda_{234}{}^{234}U}\right)(1-e^{-\lambda_{230}t}) + \left(\frac{\lambda_{234}{}^{234}U-\lambda_{238}{}^{238}U}{\lambda_{234}{}^{234}U}\right)\left(\frac{\lambda_{230}}{\lambda_{230}-\lambda_{234}}\right)(1-e^{-(\lambda_{230}-\lambda_{234})t})\right]$$

To solve these equations, assumptions about the initial content in "detrital" or "contaminating" fractions of ^{231}Pa and ^{230}Th, respectively, must then be made. In literature, this issue is often discussed and approaches vary from one study to another since the properties of detrital particles are site specific. In recent years, though, a purely analytical approach has been used to document the composition of the authigenic fraction, in samples, vs. that of the "detrital" fraction (Fig. 7). The basic assumption in this approach is that the detrital fraction is "labeled" by the presence of ^{232}Th. This isotope should indeed be practically absent in the authigenic fraction of minerals formed in aquatic environments due to its low solubility, in comparison to that of U. On the contrary, any clay fraction contaminating a carbonate will bring in significant amounts of ^{232}Th. In such cases, the analysis of several samples from the same setting but with variable amounts of detrital vs. authigenic fractions should permit to define a mixing line, in the diagram below, thus to extrapolate this line to values representing the isotopic composition of the authigenic end-member, and through equation [5] and/or [1], to calculate the age "t" of this authigenic fraction, as well as to control any later diagenetic chemical mobility of U when combining both equations. As may be understood, it is assumed that detrital fractions in all samples show similar U-series disequilibria and U/Th ratios.

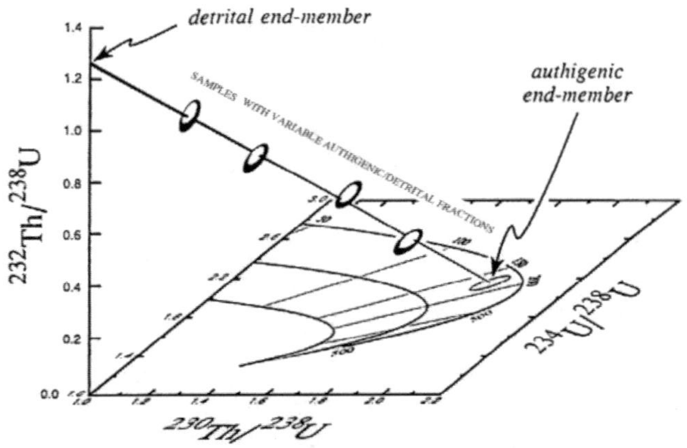

Figure 7: Mixing line between an authigenic component (the mineral to be dated) and a detrital fraction "labeled" by its ^{232}Th-content (from [11]). Note that errors bars are replaced by ellipsoids defined by errors for each datum point along each axis. The plane defined by the x- and y-axis is an abaca plot almost similar to that of Figure 6 above, but for the activity ratio $^{230}Th/^{238}U$ instead of $^{230}Th/^{234}U$.

When the detrital fraction does not carry in any ^{232}Th, this approach does not work, neither when the detrital fraction shows large amplitude heterogeneities between samples, either due to variable grain size or sources.

7. Application to the dating of biogenic carbonates

As briefly evoked above, biogenic minerals rarely behave as perfect closed systems and some diagenetic mobility of U is often observed. In biogenic minerals containing high amounts of authigenic uranium, this diagenetic mobility may result in negligible biases on age estimates. Unfortunately, most biogenic minerals show very small amounts of truly authigenic uranium as illustrated in Table 3.

Type of sample	Sampling site	U-content in ppb
Diatoms (opal)	Reykjanes Ridge core	1.016±0.047
Fish otholiths (calcite)	California coast	0.339±0.003
Foraminifers (calcite)	Greenland Rise core	4.92±0.01
Gastropods- marine (aragonite)	Chilean coast	7.43±0.03
Gastropods - terrestrial (aragonite)	Lanzarote island	1.186±0.006
Pelecypods - marine (aragonite)	Lanzarote island	29.9±0.01
Cirripeds (calcite)	Peruvian coast	18.20±0.06
Gorgonians (calcite)	NW Atlantic	< 100
Scleractinians (aragonite)	Cabo Verde	> 2000

Table 3: Minimum U-content observed in modern or fossil samples analyzed at GEOTOP (B. Ghaleb).

The only organisms showing significant amounts of U in their skeleton (~ 3 ppm) are the aragonitic scleractinian corals. Detailed studies by Robinson et al. [12] indicate that this uranium is linked to the aragonite lattice, whereas in most other biogenic minerals investigated, U seems mostly present between crystals and along organic linings, or result from minute contamination [e.g., 13, 14]. The scleractinian corals depict a U/Ca temperature-dependent partition coefficient of ~ 1 with seawater. The negative relationship seems nearly linear at ambient temperature as illustrated in Figure 8, below.

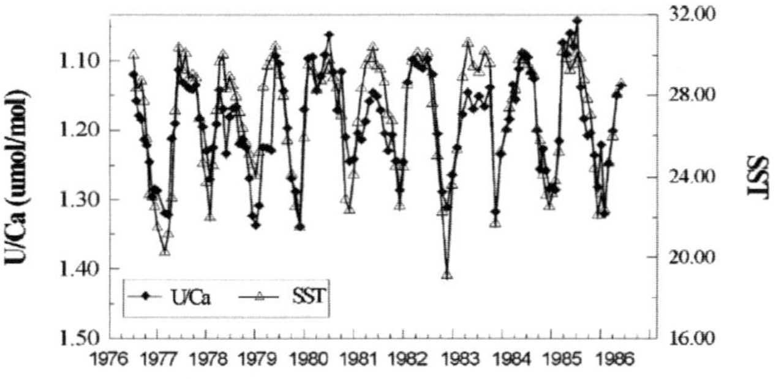

Figure 8: U/Ca molar ratio vs. surface temperature in a recent *Porites* specimen from the South China Sea (from [15]).

Nonetheless, even with such a relatively high authigenic U-buffer, U-series ages in fossil scleractinians corals are frequently biased through either some diagenetic U-mobility or discrete ^{234}Th-^{230}Th enrichment process (e.g., [16]), as illustrated Figure 9 below.

Cementation of the embedding sediment (e.g., in beach-rocks), low Eh (mud) and arid conditions illustrate such favorable settings for reduced late diagenetic U-fluxes, as documented in several studies (e.g., [17]). This latter reference provides a good illustration of the decreasing but differential diagenetic U-fluxes, through time, linked to cementation of samples with distinct relative porosities (Fig. 10, below).

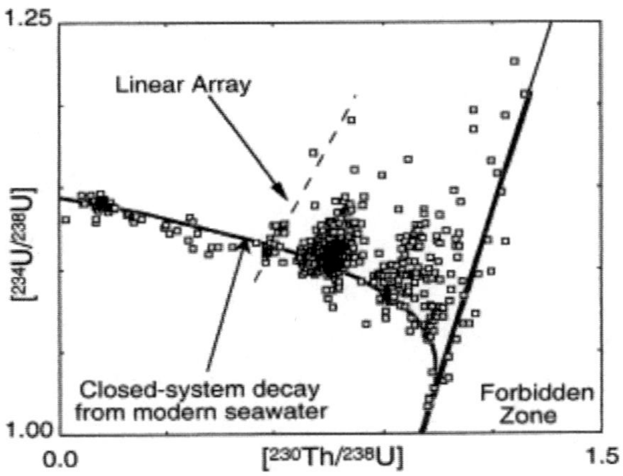

Figure 9: Systematics of U-Th activity ratios in fossil corals (data from literature compiled by Thompson et al. [6]; the figure is extracted from the same paper). Most scleractinian corals from pre-Holocene interglacials plot over the pseudo-isochron line, thus indicating an open system.

In the example illustrated Figure 10, assuming that the diagenetic U-uptake process is a continuous process described by a simple algorithmic equation, a better assessment of the mean age of this early diagenetic U-uptake phase can be obtained using a simple correction model.

Figure 10: U content and mean ^{230}Th-age of the U uptaken by gastropod shells from strongly cemented beach-rocks of Last Interglacial and mid-Holocene ages, compared with modern specimens (Lanzarote Island, [17]). Data indicate an important, early diagenetic U-uptake phase, followed by reduced U-uptake rates in later diagenetic stages, especially in the less porous shells of *Patella* sp. vs. *Thais* sp. Note the impact of these diagenetic fluxes on the apparent ^{230}Th-

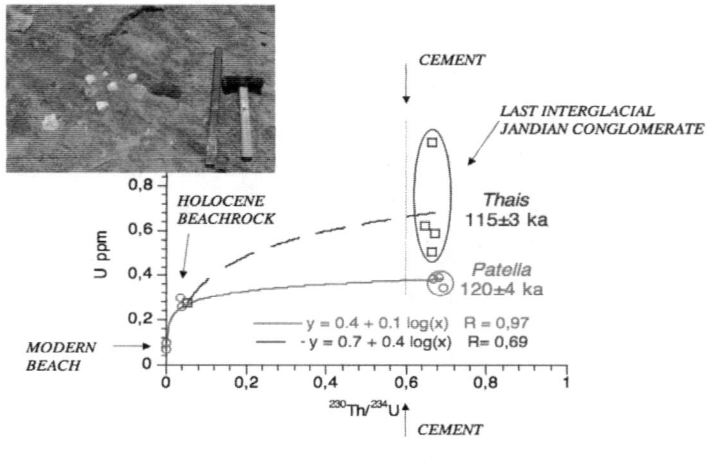

ages of the uranium uptaken by the Last Interglacial specimens.
Worth of mention is the fact that in the examples above, all fossils are aragonitic. Thus the incorporation of U in the crystal lattice is not a function of the mineral (i.e., calcite vs. aragonite, for example), but a biologically driven process linked to the mode of aragonite precipitation in the skeleton of scleractinians, nearly under kinetic equilibrium conditions with ambient water, whereas all other organisms precipitating aragonite for their shell are practically deprived of any authigenic U (Table 3).

Aragonite is unstable at low temperature and is normally replaced by calcite through diagenetic process that could also be responsible for some U and/or daughter products relocation. As a consequence, several authors use *a contrario* the mineralogical integrity of scleractinian fossils to ascertain the closure of the radioactive system (e.g., [18]). Based on high resolution X-Ray examination of samples, all specimens showing minute percentages of calcite (usually > 1%) are discarded. This method raises a few issues. Some are methodological and concern for example the sensitivity vs. precision of X-Ray measurements of very small quantities of calcite in carbonate samples (e.g., [19]). Others concern the lack of unequivocal linkage of mineral instability with the slow-rate, Th-enrichment processes observed in corals. Furthermore, if the epigenetic transformation of aragonite into calcite likely leads to redistribution of U and/or daughter elements, in the fossil specimens, between secondary minerals, the impact on the overall U-series age estimates is not necessarily obvious, as in the example below (Fig. 11).

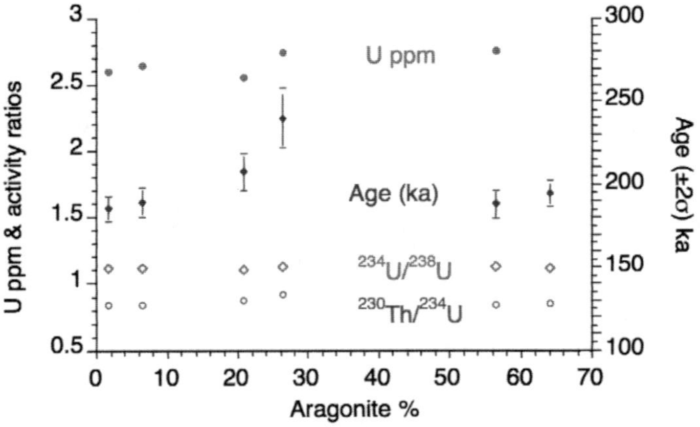

Figure 11: U-series systematics and ^{230}Th-ages in *Cladocora caespitosa* colonies (scleractinian coral) from raised marine deposits at La Marina, SE Spain (from [20]). These specimens show a very significant transformation of their skeleton into calcite, but without any clear impact on U-series systematics and ^{230}Th-ages

Aside from the "historical" U-series dating method, discussed above, attempts at using shorter-lived isotopes (e.g., ^{226}Ra and ^{210}Pb), in excess vs. their parent element, for the dating of recent biogenic minerals and organic remains, have been made with variable success. The problem, when using such methods, is to estimate the original excess of either ^{226}Ra or ^{210}Pb or of both, that may vary in time and space. More details are given in another chapter of the present volume with respect to these isotopes and their potential use. A curious, but apparently robust example of such dating, is given in the example below.

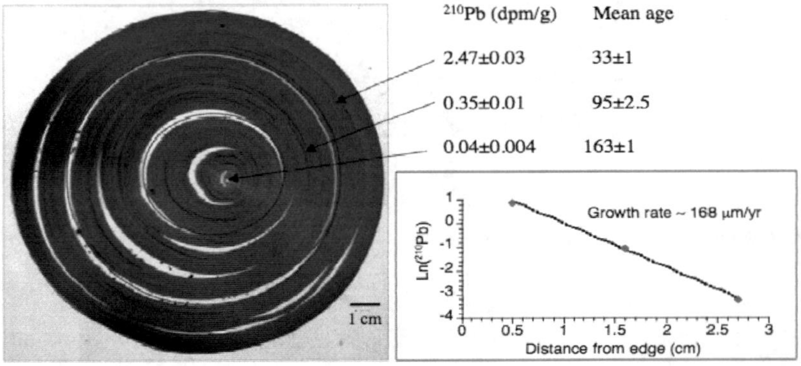

Figure 12: ^{210}Pb data in a black coral stem (cf. Anthipatharians). Using a simple model based on a constant initial ^{210}Pb-content, growth rates can be estimated (data B. Ghaleb; specimen collected off Florida by M. Risk).

8. Conclusions

The "U-series disequilibrium method" has provided important information on the ocean dynamics (cf. [21, 22]). It has also provided key data about the age of raised marine terraces linked to high interglacial sea-levels and helped validating the astronomical theory of climate (e.g., [23]). It has also been essential in the documenting of sea-level changes during the last glacial cycle in particular (e.g., [24]), as well as on the calibration of the radiocarbon time scale (e.g., [25]). However, in-depth examination of U-series systematics in fossil biogenic minerals indicate that U-mobility and other complex diagenetic processes involving daughter products, not to mention the quasi-absence of authigenic U in most of these minerals, limit the usefulness of this method for the precise dating of such biological remains.

When closed systems are maintained through time, as they often do in speleothems for example, these isotopes are critical to reconstruct paleoclimate time series. However, even in such settings, diagenetic changes with redistribution of U between secondary minerals may also result in chronological biases [26].

Open systems seem thus to be quite frequent. They do not necessarily prevent the setting of relatively reliable chronologies, especially when a high temporal resolution is not required. Flux-corrected age-models have often been proposed, based on various approaches (e.g., [6, 20]). Frequently, the diagenetic uptake of uranium occurs rapidly (cemented beach-rocks, low Eh-settings, arid zones). In such cases, U-series ages provide minimum ages for the embedding geological unit. Thus, one must conclude that if the U-series isotopes provide a unique "tool-box" for geochemical studies and the dating of recent Earth processes, they often require exhaustive investigations in given geological settings to ascertain their behavior though time, thus, proper age and flux estimates based on their radioactive decay properties.

9. References

[1] Ivanovich M and Harmon R S 1992 *Uranium-Series Disequilibrium* Clarendon Press Oxford UK
[2] Bourdon B, Henderson GM, Lundstrom CC and Turner S (eds) 2003 *Rev. Mineral. Geochem.* **52** 656 p
[3] Ghaleb B 2009 *Overview of the methods for the measurement and interpretation of short-lived radioisotopes and their limits* Present volume
[4] Dickin A.P. 2005 *Radiogenic Isotope Geology* Cambridge University Press
[5] Deschamps P, Hillaire-Marcel C, Michelot J-L, Doucelance R et al. 2004 *Hydrol. Earth System Sci.* **8** 35
[6] Thompson W G, Spiegelman M W, Goldstein S L and Speed R C 2003 *Earth Planet. Sci. Lett.* **210** 365
[7] Dunk R M, Mills R A and Jenkins W J 2002 *Chemical Geol.* **190** 45.
[8] Gariépy C, Ghaleb G, Hillaire-Marcel C, Mucci A and Vallières S 1994 *Can. J. Earth Sci.* **31** 28
[9] Chkir N and Zouari K 2007 *Environ. Geol.* **53** 677
[10] Delanghe E, Bard E and Hamelin B 2002 *Marine Chem.* **80** 79
[11] Ludwig K R and Titterington D M 1994 *Geochim. Cosmochim. Acta* **58** 5031
[12] Robinson L F, Adkins J F, Fernandez D P, Burnett D S et al. 2006 *Geochem. Geophys. Geosystems* **7** Q05022, doi:10.1029/2005GC001138
[13] Kaufman A, Ghaleb B, Wehmiller J F and Hillaire-Marcel C 1996 *Geochim. Cosmochim. Acta* **19** 3735
[14] Labonne M and Hillaire-Marcel C 2000 *Geochim. Cosmochim. Acta* **64** 1523
[15] Wei G J, Sun M, Li X H and Nie B F 2000 *Palaeogeogr. Palaeoclim. Palaeoecol.* **162** 59
[16] Edwards R L, Edwards R L, Gallup C D and Cheng H 2003 *Rev. Mineral. Geochem.* **52** 363
[17] Zazo C, Goy J-L, Hillaire-Marcel C, Ghaleb B et al. 2002 *Quaternary Sci. Rev.* **21** 2019
[18] Fairbanks R G, Mortlock R A, Chiu T-C, Cao L et al. 2005 *Quaternary Sci. Rev.* **24** 1781

[19] Reimer P J, Baillie, M G L, Bard E, Beck J W et al. 2006 *Quaternary Sci. Rev.* **25** 855
[20] Goy J L, Hillaire-Marcel C, Zazo C, Ghaleb B et al. 2006 *Geodynamica Acta* **19** 409
[21] Henderson G M and Anderson R F 2003 *Rev. Mineral. Geochem.* **52** 493
[22] Cochran J K and Masqué P 2003 *Rev. Mineral. Geochem.* **52** 461
[23] Chappell J and Shackleton N J 1986 *Nature* **324** 137
[24] Fairbanks R G 1989 *Nature* **342** 637
[25] Bard E, Hamelin B, Fairbanks RG and Zindler A 1990 *Nature* **345** 405
[26] Ortega R, Maire R, Devès G and Quinif Y 2005 *Earth Planet. Sci. Lett.* **237** 911

IOP Conf. Series: Earth and Environmental Science **5** (2009) 012009 doi:10.1088/1755-1307/5/1/012009

Black Carbon and other refractory forms in recent sediments from the Gulf of Cadiz, Spain

F J González-Vila[1], J M de la Rosa and J A González-Pérez
Instituto de Recursos Naturales y Agrobiología de Sevilla (IRNAS)-C.S.I.C.
Av, Reina Mercedes., 10, 41012-Seville, Spain.

E-mail: fjgon@irnase.csic.es

Abstract. Black carbon (BC) produced by incomplete combustion of fossil fuels and vegetation is a ubiquitous form of refractory organic matter (ROM) widely distributed in the environment. The stable non-reactive, mainly aromatic, BC has received special interest in recent years as a possible carbon sink in soils and sediments. At present, several methods have been used, with applications for soil and sediments, although no single method has been accepted to detect BC forms, and therefore providing the analytical data required to estimate BC contribution to terrestrial and sedimentary carbon fluxes and the global carbon cycle. This study considers the potential of the combined use of analytical pyrolysis (Py-GC/MS) and solid state ^{13}C NMR to determine the presence of BC and to characterize the refractory organic matter in a set of recent marine sediments from the Southwest Atlantic coast of Spain (Gulf of Cadiz). The content of BC-like material found in sediments (3.0 - 16.5 % of total organic carbon) was in the range previously reported for this kind of samples and decreases from coast to offshore. Solid state CP-MAS ^{13}C NMR and Py-GC/MS analysis of the isolated ROM revealed a conspicuous aliphatic domain remained in the samples after chlorite oxidation. Our results suggests that, in recent sediments from fluvial and marine environments, not only BC and refractory carbon forms with aromatic nature may be sequestered in a stable form; it also seems probable that the ROM pool contains a significant amount of C immobilized in the form of alkylic compounds.

1. Introduction

Black carbon (BC), soot, elemental carbon and charcoal are different terms used to describe a biologically refractory form of organic matter that remains as a residue from incomplete combustion processes. Although BC is chemically heterogeneous, a major aromatic domain is generally accepted for this material [1]. BC is ubiquitous in the atmosphere, ice, soils and sediments due to its widespread production, and its stability and inertness in the environment [2]. Moreover, it has been used as an indicator for the occurrence of forest fires and fossil fuel emissions [3]. Thus BC can represent a significant carbon sink, because it represents carbon that is transferred from the rapidly cycling biological-atmosphere carbon cycle to the long-term geological carbon cycle [4]. Considering the fact that several model calculations suggest that more than 90 % of BC deposition occurs on continental shelves, undeniably, BC can

[1] To whom any correspondence should be addressed.

IOP Conf. Series: Earth and Environmental Science **5** (2009) 012009 doi:10.1088/1755-1307/5/1/012009

contribute greatly to the organic matter being buried in marine sediments [5], and we seek to test this concept in sediments of the Gulf of Cadiz.

However, despite its importance to carbon-cycling dynamics, BC remains a component of that large fraction of OC that is classified as biochemically uncharacterized, the main reason for this is that BC is not a pure substance but rather a continuum of condensed aromatic structures with different physicochemical properties [6] and its isolation and measurement is subject to serious experimental constraints [7]. In Figure 1, taken from [8], BC materials are shown as a continuum of combustion products, "ranging from slightly charred, degradable biomass to highly condensed, refractory soot".

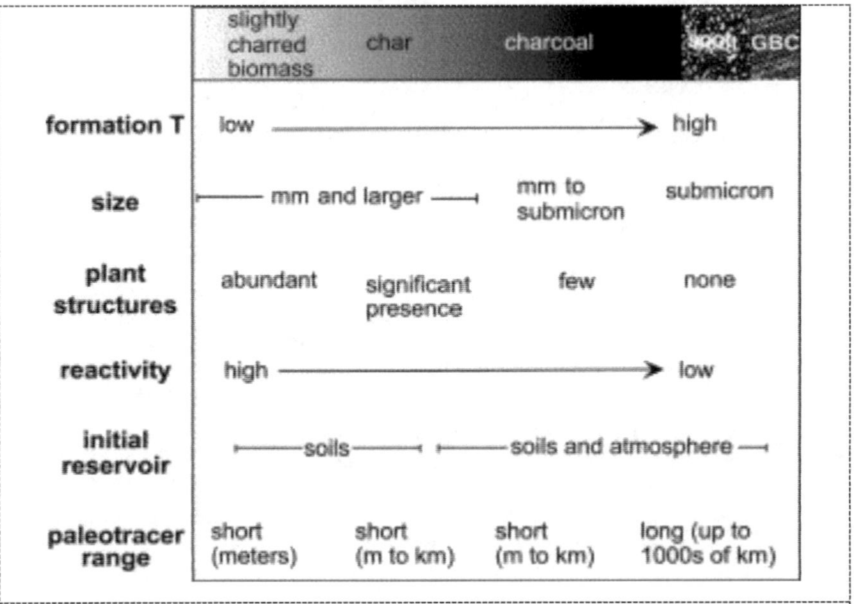

Figure 1. The Black Carbon "*continuum*" (from Massiello [8]).

Various protocols exist for the identification and quantification of BC [9, 10, 11] utilizing for example chemical, optical or thermal methods to distinguish BC from other organic and inorganic carbon in combination with elemental analysis or spectroscopy. For a comparative analysis of BC in soils and sediments it has been shown, however, that amounts of BC may vary by up to two orders of magnitude depending on the applied method [9, 12]. Creation of BC artifacts, removal of BC during pre-treatment and errors in the final estimation of BC after pre-treatment have been discussed as reasons for the high variability among methods [13]. Depending on their selectivity, different methods probably isolate different fractions of the BC continuum [14].

Seeking for an accurate analytical protocol for quantifying BC stored in different environmental complex matrices, Simpson & Hatcher (2004) [15] developed a procedure based in the removal of mineral fractions by treatment with hydrofluoric acid, followed by chemical oxidation with sodium chlorite to remove non-BC aromatic compounds (mainly lignin derived). Therefore, this study considers the potential of the combined use of analytical pyrolysis (Py-GC/MS) and solid state ^{13}C NMR to determine the presence of BC and to characterize the refractory organic matter in a set of recent marine sediments from the Southwest Atlantic coast of Spain (Gulf of Cadiz)

2. Materials and methods

In the current study, we applied the Simpson & Hatcher procedure to study the molecular features of ROM and to quantify BC content in recent sediment samples taken from the Southwest Atlantic coast of Spain (Gulf of Cadiz), a site of geological and environmental interest [16 and references therein]. Surface sediment samples (0 to 20 cm depth) were collected of-shore and in the estuaries of the Guadiana, Tinto, Odiel and Piedras rivers (Fig. 2). The area encompasses approximately 90000 km^2 of considerable deposition. Details of equipment and methods applied as well as sampling details and location are described in detail in de la Rosa et al. (2007) [17].

The quantification of the BC involves the use of solid-state CP-MAS ^{13}C NMR and the integration of the aromatic regions of the spectra after HCl/HF and $NaClO_2$ treatments. However, aromatic lignin signals overlap with those of BC and the removal of lignin is necessary for estimating BC contents. Analytical pyrolysis (Py-GC/MS) in the presence of tetramethylammonium hydroxyde (TMAH) was used to investigate the molecular composition of ROM as to test for the presence of lignin derived moieties after chlorite oxidation. This method has been found useful for the detection of lignin biomarkers [18].

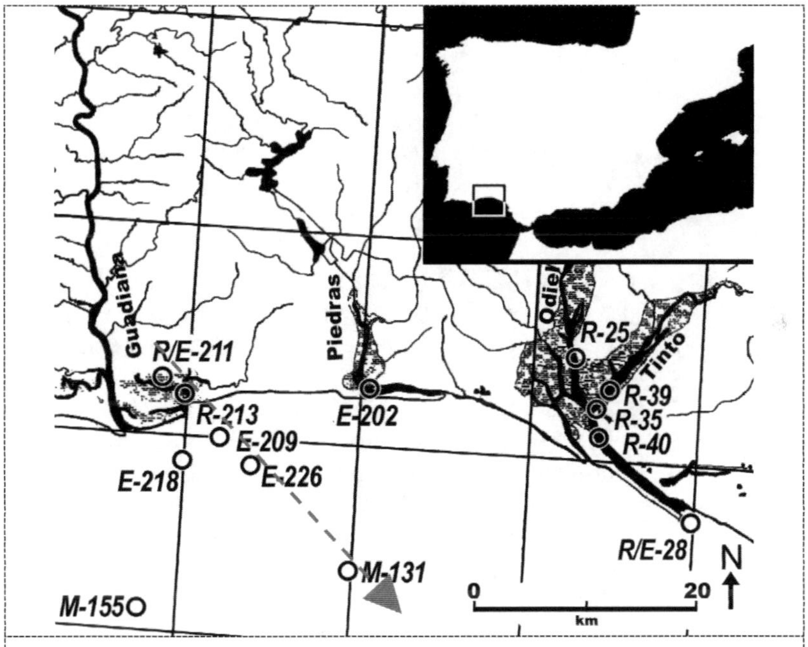

Figure 2. The sampling area. Red arrow indicates the selected transect. Sample code corresponds to location as follows R: fluvial; E: estuarine; M: marine

3. Results and discussion

After chlorite oxidation, the analytical pyrolysis in presence of tetramethylamonium hydroxide (TMAH Py-GC/MS) confirmed an evident reduction of lignin derived moieties (methoxyphenols) that could interfere with NMR aromatic signal overestimating BC content.

The content of BC as estimated by CP MAS ^{13}CNMR is shown in Table 1. BC-like material values for sample R-25 are very high (45.7 % TOC). This sample is located in the Odiel River and has conditions for accumulation of sediments with charred material from the several forests located upstream. On the other hand, this place is also affected by heavy ships traffic due to the nearby port of Huelva. The rest of the values obtained are in the usual ranges reported for BC-like materials in marine sediments [4, 19], ranging from the 3.0% BC/TOC obtained in the E-218 sample located at the end of the Guadiana estuary to the 16.5% in the R/E-211 sample (located in the mouth of the Guadiana river).

When analyzing a selection of samples that forms transect from NW to SE (Fig. 2), it is observed that BC content is independent of the TOC in the sample and that BC decreases from coast to offshore sediments (Fig. 3).

Table 1. Sediment BC content as estimated by the Simpson & Hatcher [15] method.

Ref.	gBC/Kg	BC/TOC %
R-25	8.09	45.7
R/E-28	0.78	4.3
R-35	0.90	8.2
R-39	0.81	7.8
R-40	0.87	7.9
E-202	0.47	4.1
E-209	0.47	6.1
R/E-211	1.76	16.5
E-213	1.87	15.3
E-218	0.52	3.0
E-226	0.59	5.0
M-131	0.64	5.5
M-155	0.67	6.1

CP MAS ^{13}C-NMR analysis also shows that after severe sample oxidation with chlorite. there are conspicuous signals at 110-60 and 45-0 ppm that corresponds to O-alkyl and C-alkyl moieties respectively (Fig. 4). This accounted for c. 60% of total NMR signal (O-alkyl 27.5 ± 2.83 and C-alkyl 31.9 ± 4.54).

Also TMAH Py-GC/MS analysis revealed a highly aliphatic nature of the ROM isolated from marine sediments, even in the sample with the highest value estimated for BC (R-25) (Fig. 5 and Table 2). The chromatograms are dominated by series of methyl and dimethyl esters of fatty acids including branched molecules. Also present are benzene and naphthalene derivatives.

Both CP MAS ^{13}C NMR spectra and TMAH Py-GC/MS chromatograms shows that after a severe oxidation of, a conspicuous lipidic and peptidic domain still remains in the sedimentary OM as recalcitrant fraction. That fraction of ROM could be considered out of the so called continuum concept to define BC material. Our results point to a strong recalcitrance of aliphatic moieties in such environments that, together with the condensed BC forms, these moieties are main components of fluvial and marine ROM fraction and that as previously suggested [10] this refractory, non-oxidable, aliphatic carbon pool, may represent a stable OM fraction with influence in C fluxes and therefore in the global biogeochemical C cycle.

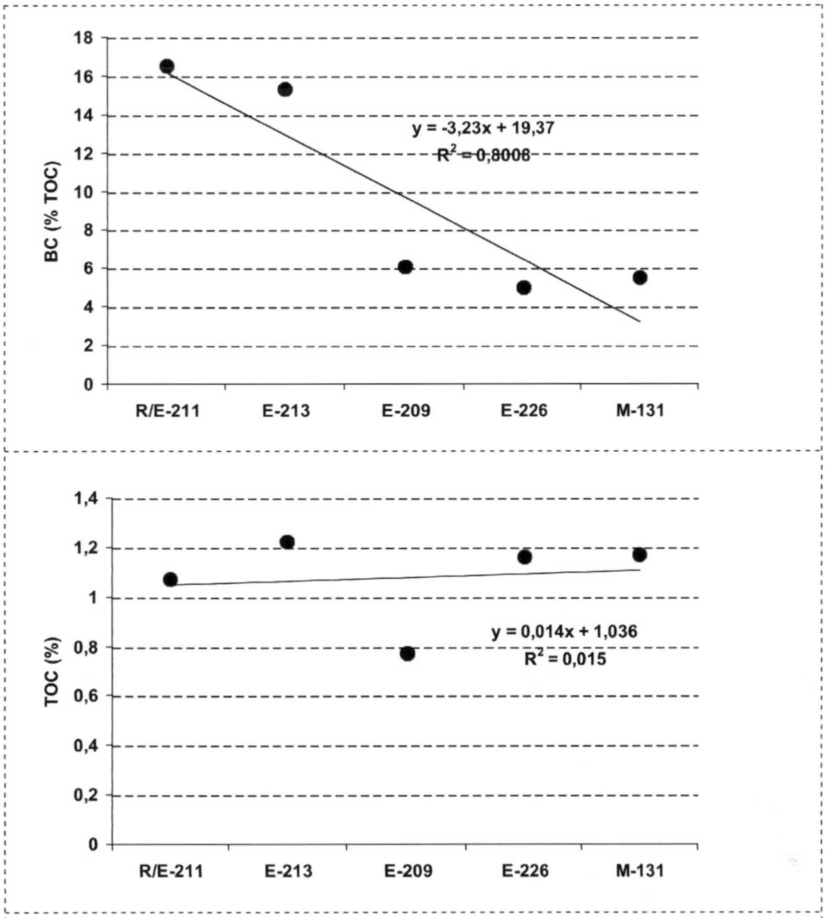

Figure 3. BC (% TOC) and TOC (%) in a transect from NW to SE Guadiana River in the gulf of Cadiz as illustrated in Fig 2..

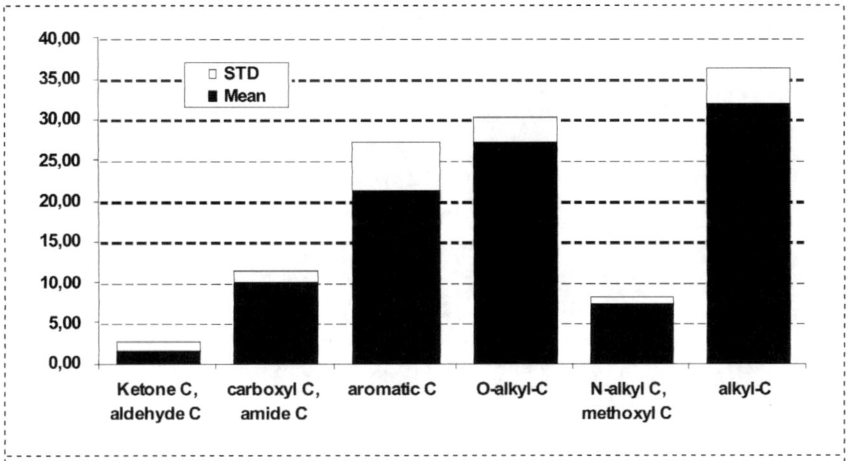

Figure 4. Solid state CP MAS ^{13}C NMR integration regions after HCl/HF and $NaClO_2$ treatments. Results are given in % and n=13.

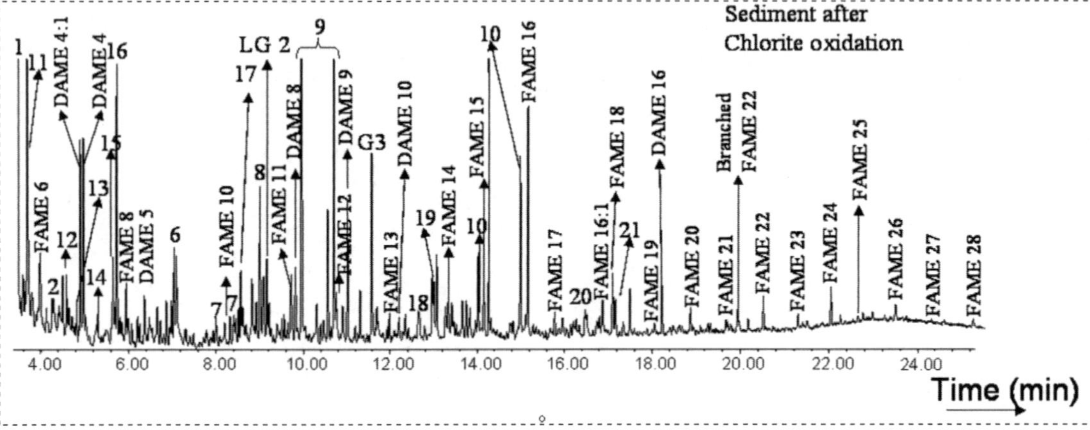

Figure 5. Example of TMAH Py-GC/MS chromatogram of sediment (R-25) after chlorite oxidation. The corresponding biomarker abbreviations are listed in Table 2. Methyl esters of fatty acids are identified as FAMEn, dimethyl diesters of fatty acids are identified as DAMEn, where "n" is the length of the carbon chain. LG: Lignin derived compound.

Table 2. Peak identification of TMAH Py-GC/MS non lignin origin released compounds. Label refers to peks in Figure 5.

Label	Non-Lignin Compounds released by Py-GC/MS
1	Dimethyl sulfone
2	N.N-dimethyl-t-butylamine
3	Glycine-N-acetyl methyl ester
4	Benzoic acid, methyl ester
5	Dimethyl, methyl fumarate
6	Benzoic acid, 3-methyl-methyl ester
7	Naphthalene, 2-methyl
8	1.3.5-Triazine-2.4.6 (1H.3H.5H)-trione-trimethyl
9	1.2-Benzenedicarboxilic acid dimethyl ester
10	1.2.4-Benzenetricarboxylic acid, trimethyl ester
11	N.N-Dimethyl-propanamide
12	Phenol
13	1.2.3-Trimethylbenzene
14	m-Cresol
15	Glicine.N(2-methoxy-2-oxoethyl)-methyl ester
16	Benzoic acid, 3-methyl-methyl ester
17	N-methylpyroglutamic acid, methyl ester
18	2-Naphthalenecarboxylic acid, methyl ester
19	2.4.6-Trimetoxybenzonitrile
20	2.6-Naphthalenedicarboxylic acid, dimethyl ester
21	1.2.4.5-Bencenetetracarboxylic acid, tetramethyl ester

Acknowledgements

We thank Ms Trinidad Verdejo for assisting us in the Py-GC/MS spectroscopy.The SpanishMinistry ofEducation provided financial support for this research (fellowship BES-2003-1900 and project CGL2006-12730-C03-C01). We would also like to thank Dr. E. López-Capel for helpful comments when reviewing the manuscript.

References

[1] Jones, T.P. & Chaloner, W.G. 1991. Fossil charcoal, its recognition and palaeoatmospheric significance. *Global and Planetary Change*, **5**: 9–50.

[2] Goldberd, E.D. 1985. Black Carbon in the Environment. Wiley-Interscience, N.Y.

[3] Bird, M.I. & Cali, J.A. 1998. A million-year record of fire in sub-Saharan Africa. *Nature*, **394**: 767 – 769.

[4] Middelburg, J.J., Nieuwenhuize, J. & Van Breugel, P. 1999. Black carbon in marine sediments. *Marine Chemistry*, **65**: 245–252.

[5] Masiello, C.A., Druffel, E.R.M. & Currie, L.A. 2002. Radiocarbon measurements of black carbon in aerosols and ocean sediments. *Geochimica et Cosmochimica Acta*, **66**: 1025–1036.

[6] Hedges, J. I., Eglinton, G., Hatcher, P.G., Kirchman, D.L., Arnosti, C., Derenne, S., Evershed, R.P., Kögel-Knabner, I., de Leeuw, J.W. & Littke, R. 2000. The molecularly-uncharacterized component of nonliving organic matter in natural environments. *Organic Geochemistry*, **31**: 945-958.

[7] Kuhlbusch, T.A.J. 1998. Black carbon and the carbon cycle. *Science*, **280**: 1903–1904

[8] Massiello, C.A. 2004. New directions in black carbon organic geochemistry. *Marine Chemistry*, **92**: 201–213.

[9] Hammes, K., Schmidt, M.W.I., Smernik, R.J., Currie, L.A.R., Ball, W.P., Nguyen, T.H., Louchouarn, P., Houel, S., Gustafsson, Ö., Elmquist, M., Cornelissen, G., Skjemstad, J.O., Masiello, C.A., Song, J., Peng, P., Mitra, S., Dunn, J.C., Hatcher, P.G., Hockaday, W.C., Smith, D.M., Hartkopf-Fröder, C., Böhmer, A., Lüer, B., Huebert, B.J., Amelung, W., Brodowski, S., Huang, L., Zhang, W., Gschwend, P.M., Flores-Cervantes, D.X., Largeau, C., Rouzaud, J.-N., Rumpel, C., Guggenberger, G., Kaiser, K., Rodionov, A., Gonzalez-Vila, F.J., Gonzalez-Perez, J.A., de la Rosa, J.M., Manning, D.A.C., López-Capél, E., Ding, L., 2007. Comparison of black carbon quantification methods using reference materials from soil, water, sediment and the atmosphere, and implications for the global carbon change. *Global Biogeochemical Cycles*. **21**: GB3016. DOI: 10.1029/2006GB002914

[10] De la Rosa, J.M., González-Pérez, J.A., Hatcher, P.G., Knicker, H., González-Vila, F.J. 2008. Determination of refractory organic matter in marine sediments by chemical oxidation, analytical pyrolysis and solid-state ^{13}C nuclear magnetic resonance spectroscopy. *European Journal of Soil Science* **59**: 430-438.

[11] Skjemstad, J.O., Taylor, J.A., Smernik. R.J. 1999. Estimation of charcoal (char) in soils. *Communications in Soil Science and Plant Analysis*. **30**: 2283-2298.

[12] Schmidt, M.W.I., Skjemstad, J.O., Czimczic, C.I., Glaser, B., Prentice K.M., Gelinas, Y. & Kuhlbusch, T.A.J. 2001. Comparative analysis of black carbon in soils. *Global Biogeochemical cycles*, **15**: 163–167.

[13] Nguyen, T.H., Brown, R.A. & Ball, W.P. 2004. An evaluation of thermal resistance as a measure of black carbon content in diesel soot, wood char, and sediment. *Organic Geochemistry*, **35**: 217-234.

[14] BC Ring-Trial. *www.geo.unizh.ch/phys/bc/ringtrial.html*

[15] Simpson, M., & Hatcher, P. G., 2004. Determination of Black Carbon in natural organic matter by chemical oxidation and solid-state ^{13}C nuclear magnetic resonance

spectroscopy. *Organic Geochemistry*, **35**: 923-935.

[16] González-Pérez, J.A., De Andrés, J.R., Clemente, L., Martín, J.A., González-Vila, F.J., 2008. Organic carbon and environmental quality of riverine and off-shore sediments from the Gulf of Cádiz, Spain. *Environmental Chemistry Letters* **6**: 41-46

[17] de la Rosa, J.M., Lopez Capel, E., Gonzalez-Vila F.J., Gonzalez-Perez J.A., Manning, D.A.C., 2008. Direct detection of black carbon in soils by Py-GC/MS, ^{13}C NMR spectroscopy and thermogravimetric techniques. *Soil Science Society of America Journal*, **72**: 258-267.

[18] del Río, J.C., & Hatcher, P.G. 1998. Analysis of aliphatic biopolymers using thermochemolysis with tetramethylammonium hydroxide (TMAH) and gas chromatography–mass spectrometry. *Organic Geochemistry*, **29**: 1441–1451.

[19] Gustafsson, Ö. & Gschwend P.M., 1997. Soot as a strong partition medium for polycyclic aromatic hydrocarbons in aquatic systems. *In*: R.P. Eganhouse, Editor, *Molecular Markers in Environmental Geochemistry*. ACS Symp. series 671, ACS, Washington, DC: 365–381 *pp*.

IOP Conf. Series: Earth and Environmental Science **5** (2009) 012010 doi:10.1088/1755-1307/5/1/012010

Density separation of combustion-derived soot and petrogenic graphitic black carbon: Quantification and isotopic characterization

Marie-Hélène Veilleux[1], Angela F. Dickens[2], Jay Brandes[3] and Yves Gélinas[1]

[1]GEOTOP, and Department of Chemistry and Biochemistry, Concordia University, 7141 Sherbrooke Street West, Montreal (Quebec), Canada, H4B 1R6

[2]Mount Holyoke College, Chemistry Department, 50 College St., South Hadley, MA 01075-6407, USA

[3]Skidaway Institute of Oceanography, 10 Ocean Science Circle, Savannah, Georgia 31411, USA

ygelinas@alcor.concordia.ca

Abstract. The black carbon continuum is composed of a series of carbon-rich components derived from combustion or metamorphism and characterized by contrasting environmental behavior and susceptibility to oxidation. In this work, we present a micro-scale density fractionation method that allows isolating the small quantities of soot-like and graphitic material usually found in natural samples. Organic carbon and $\delta^{13}C$ mass balance calculations were used to quantify the relative contributions of the two fractions to thermally-stable organic matter from a series of aquatic sediments. Varying proportions of soot-like and graphitic material were found in these samples, with large variations in $\delta^{13}C$ signatures suggesting important differences in their origin and/or dynamics in the environment.

1. Introduction

Marine sediments are the ultimate sink for natural organic matter, and the key gateway between the global organic carbon cycle at the Earth's surface and the slowly cycling carbon pool, whose turnover is tightly coupled to subduction and uplift of the crust. Sedimentary organic matter (SOM) comprises a myriad of highly altered organic compounds sorbed onto and/or entrapped within the mineral matrix of the sediment, forming a dynamic mineral-organic soup of daunting complexity. Extensive (photo/bio)chemical alteration, resistance to hydrolysis, poor solubility in aqueous and non-polar solvents, and physical protection have all been linked to the fact that the majority of the SOM pool (>80% in most cases) is still uncharacterized at the molecular level using conventional chromatographic and spectroscopic techniques [1]. It is mostly this lack of molecular detail on the chemical makeup of SOM that has delayed the construction of a comprehensive map detailing the origins, reaction pathways and ultimate fate of individual organic compounds in these settings.

Our continuing inability to molecularly characterize the organic remains of living organisms accumulating in sediments has fuelled the significant advances in bulk characterization approaches and techniques that have dominated the field of Organic Geochemistry in recent years. Technical development and elegant new approaches exploiting spectroscopic properties of the SOM have now been successfully applied (e.g., absorbance/fluorescence [2]; NMR [3-5]; X-ray spectromicroscopy [6,7]), with the overarching aim of narrowing the gap between the molecular and bulk levels of information concealed within the sedimentary organic soup.

In parallel to these technological advances, strategies based on the fractionation of SOM into operationally-defined pools of compounds exhibiting contrasting chemical/thermal reactivities have also been revisited [8-9]. Using wet chemistry approaches, these strategies are designed to reduce the level of complexity of bulk SOM; they rely on the assumption that each isolated pool is predominantly composed of organic compounds with similar physicochemical properties. While these approaches generate useful and much needed information at the bulk level, their lack of molecular specificity severely limits our ability to anchor this bulk information to mechanistically explicit molecular models. What has been traditionally labelled "black carbon" is one of the best examples of such operationally-defined and/or chemically fractionated organic carbon pools.

Black carbon (BC) is a generic term that encompasses a continuum of carbon- and aromatic-rich organic compounds comprising products generated through the combustion of biomass and fossil fuels, or as a result of sedimentary metamorphism (Figure 1) [10].

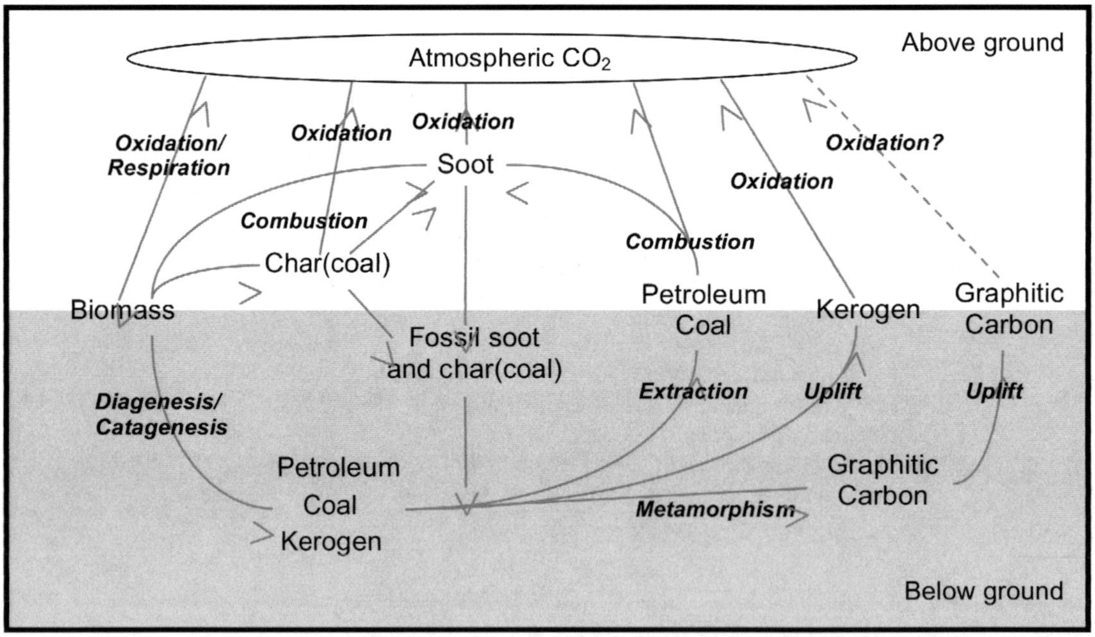

Figure 1: The black carbon cycle. Black carbon components are divided between an above-ground, combustion driven, quickly cycling carbon pool (char, charcoal and soot), and a predominantly below-ground, petrogenic, slowly cycling pool (coal, fossil soot and charcoal, graphitic carbon and pure graphite). Full arrows indicate confirmed, but not yet quantified, fluxes while the serrated arrow is a flux that remains to be confirmed. Components traditionally included in the black carbon continuum are char, charcoal, coal, and soot; the addition of graphitic carbon and pure graphite to the list is debated.

These compounds range from slightly charred plant remains to highly condensed graphite, and include combustion-derived materials (char, charcoal and soot), as well as materials of petrogenic origin (coal, graphitized carbon and pure graphite). The physical and chemical properties of each class of compounds, and thus their resistance to oxidation, are determined in large part by their formation pathways. Compounds formed in the solid state such as char/charcoal (from the incomplete combustion of plant materials), and coal (product of the maturation of peat remains) retain some of the original properties of their source material. Soot, on the other hand, is formed through the high temperature condensation of hot gases emanating from solid and liquid fuels during combustion; it retains very few of the physical and chemical properties of its source material. The chemical composition and structure of the different components vary in a gradual way along the BC continuum, which prevents the definition of clear boundaries between the different classes of compounds. This absence of discrete definitions, coupled to the highly insoluble macromolecular nature of BC compounds, is responsible for the dearth of analytical methods that quantitatively measure or isolate any single class of compounds within the BC continuum. Research projects targeting specific BC compounds or the whole BC continuum are thus analytically extremely challenging [10, 11, 12].

One of the more recent methods for analyzing BC in soils and sediments exploits the contrasting thermal resistance of the different classes of compounds by combusting the carbonate-free samples at 375 °C in an O_2-saturated atmosphere [13], based on the early work of Cachier and coworkers [14]. The method targets soot, one of the most thermally-stable BC components, although its specificity (accuracy) has been questioned, particularly when applied to nitrogen-rich samples [15-16]. Combustion of N-rich samples leads to the formation of condensation products when macromolecular biochemicals are depolymerised and volatilized at high temperatures [17]. Condensation products sometimes translate into severe overestimations of the real soot concentrations [15]. While this potential artefact is minimized when analyzing N-depleted samples (such as freshwater sediments and soils), and by optimising O_2 access to the sample during combustion [16,18], the measurement of soot in N-rich samples (such as marine sediments containing fresh organic matter) is still problematic, particularly when their soot content is low. This problem has led to the development of alternative methods in which the potentially interfering biochemicals are chemically removed before thermal treatment [15]. Still, a recent natural BC quantification intercomparison has shown all methods are prone to their drawbacks and no single approach can be said to specifically and accurately quantify BC [19].

Despite the impressive efforts deployed in the last five years to identify and circumvent the potential biases affecting each BC method, another potential source of ambiguity, affecting mostly carbon mass balance and thermal based BC determination methods, has been overlooked: petrogenic graphitic carbon [20-21]. Graphitic carbon is formed in the crust at high pressure and temperature during metamorphism or through direct precipitation from high-CO_2 bearing hydrothermal fluids (Figure 1) [7]. Upon uplifting of the crust, graphitic carbon becomes exposed at the surface of the Earth and is eventually deposited into soils and sediments through erosion and runoff. Graphitic carbon is highly resistant to thermal and chemical oxidation and is thus not altered or removed during the chemical and thermal pre-treatments used in many BC determination methods. Depending on the relative concentration of soot and graphitic carbon in a sample, this bias can be important: As shown in ref. [21], the graphitic carbon concentration can be much higher than that of soot in some samples.

The issue is important because soot and graphitic carbon behave very differently in natural environments. While small soot particles can rapidly be oxidized by exposure to UV and activated oxygen species [22], the highly aromatic crystalline structure of graphitic carbon makes this component especially resistant to (bio/photo)chemical degradation. The contrasting size and density of soot and graphitic carbon particles also lead to very different dispersion pathways in the environment [21]. There is thus a need to develop a method allowing the quantification and fractionation of soot and graphitic carbon into discrete components. Dickens et al. [21] used a large volume heavy-liquid floatation approach to fractionate a large quantity of bulk sediments into fractions of varying density. Each separated fraction was then analyzed for their soot and graphitic carbon content (labeled the GBC fraction) using the chemical and thermal approach described in ref. [15]. However, that work was based on the unconfirmed assumption that soot particles are not associated with high density particles such as minerals, which, if not true, would lead to an overestimation of graphitic carbon, and an underestimation of the soot contribution to GBC.

To verify the above assumption, a quantitative micro-scale density fractionation of mineral-free soot and graphitic carbon present in GBC isolated from marine and freshwater sediments was optimized. Density fractionation using heavy liquid (sodium polytungstate) floatation was chosen based on the original work of Dickens et al. [21]. The method was validated using several standards and natural samples through elemental and isotopic mass balance calculations.

2. Materials and Methods

2.1. *Samples.* The samples used in this work encompass a wide range of aquatic environments. The Stillaguamish River (48°05'N, 121°45'W) sample is an organic-poor (0.142 wt% OC) surface sediment collected in a river draining the foothills of the Cascades Mountains about 75 km north of Seattle, WA, USA [7,20]. The Saguenay sample is a surface sediment (0-5 cm, 2.30 wt% OC) collected in the deepest part of the Saguenay Fjord, Québec, Canada (48°15.7"N, 70°09.5'W). The Lake Washington (47°37'N, 122°15.8'W) sample is a composite (0-25 cm, 4.69 wt% OC) sediment collected in a large urbanised lake about one km downwind from downtown Seattle, WA, USA [23]. The Buffalo River sediment is the standard reference material SRM 2704 (NIST, Gaithersburg, MD, USA; 2.43 wt% OC) collected downtown Buffalo, NY, USA. All samples were quickly frozen or freeze-dried upon collection.

2.2. *Isolation of the GBC fraction.* The GBC fraction was isolated through chemical (demineralization and hydrolysis) and thermal treatments of a sample, as detailed in ref. [15]. Briefly, the procedure involves the sequential removal of (i) carbonates and amorphous oxides with 1N hydrochloric acid (shaking for 6 hours at room temperature), (ii) silicate minerals by treating the sample between 2 and 5 times with a mixture of 10% hydrofluoric acid and 1N hydrochloric acid (12 hours at room temperature), (iii) hydrolysable organic matter composed predominantly of polysaccharides using three treatments of 2N, 4N and 6N, respectively, of trifluoroacetic acid (12 hours at 95 °C under N_2), (iv) hydrolysable organic matter composed predominantly of peptides and proteins with 6N hydrochloric acid (12 hours at 110 °C), and, (v) thermally labile non-hydrolyzable organic matter by combustion at 375 °C in an O_2-saturated atmosphere for 24 hours.

2.3. *Density fractionation.* The sodium polytungstate (SPT) heavy liquid floatation procedure used in this work was adapted from the method described by Dickens et al. [21]. The procedure was reduced to the micro-scale level to allow the estimation of the usually very small quantities of combustion-derived soot and petrogenic graphitic carbon in GBC isolated from natural sediments. Fractionation was carried out using a solution of SPT adjusted to a density of 1.8 g mL^{-1} with distilled water. This density roughly corresponds to the middle of the range separating soot (1.2 – 1.8 g mL^{-1} [24,25]) and graphite (2.1 – 2.3 g mL^{-1} [26]). To ensure that optimal conditions were used for the separation of soot and graphitic carbon from sediments, the method was first validated with n-hexane soot and graphite standards (Table 1). Because the $\delta^{13}C$ signature of these two BC components is slightly different, separation efficiency was evaluated by both isotope and mass balance.

Density fractionation was carried out in 1.5-mL silanized eppendorf tubes (Fisher Scientific). One mL of the 1.8 g mL^{-1} SPT solution was added to the GBC isolate. The tube was then sonicated for 30 minutes and ultracentrifuged at 12,000 rpm for 15 minutes. The soot-containing supernatant was then removed and filtered on 25-mm Isopore® polycarbonate filters (0.2 μm nominal pore size). This first step was repeated twice to ensure complete separation of the light material from the heavier fraction. The filter was then rinsed with ~5 mL of distilled water to remove residual polytungstate salts. The soot fraction was then very gently scraped off the filter using a stainless steel spatula and transferred into a small pre-weighed aluminum cup for drying and weighing. The choice of the polycarbonate filters was dictated by their smooth glass-like surface with uniform cylindrical pores (as opposed to the GFF filters made out of borosilicate glass fibers woven into a porous matrix). Because the surface of the filter is hydrophilic, soot and graphitic carbon do not sorb to the filter and can be quantitatively recovered by gentle scraping followed by rinsing with water (and methanol, if necessary, although it was not the case in this work). Organic carbon contamination from the polycarbonate filters was also verified and found to be negligible. The same filtration approach was used to recover and rinse the high density graphitic carbon residue from the eppendorf tube. Both fractions were accurately weighed using a Mettler-Toledo MX-5 analytical microbalance (precision ± 1 μg).

Elemental and stable isotope analysis: The organic carbon concentrations and stable carbon isotope compositions (vs. VPBD) were measured on a Europa elemental analyser coupled to an Isoprime continuous flow isotope-ratio monitoring mass spectrometer (EA-IRMS, Elementar Americas, NJ, USA). To ensure optimal combustion efficiency of the thermally stable GBC components in the EA, the combustion temperature was set to the maximum working value of 1100 °C, and the amount of O_2 co-injected with the sample was doubled.

3. Results and discussion

3.1. *The GBC fraction.* GBC is an operationally defined fraction of the black carbon continuum. It comprises the most refractory components, i.e., soot and graphitic carbon, which are isolated by combusting a demineralised and hydrolysed sample at 375 °C. The choice of the combustion temperature was made after testing for the thermal resistance of an array of organic compounds, including pure components from the BC continuum [13]. The organic residue remaining following this series of increasingly harsh treatments predominantly includes soot and petrogenic graphitic carbon [20], although a small fraction of the most recalcitrant type of coal (anthracite, if contained in the sample), might also be present. It has also been suggested that the soot fraction might not be quantitatively recovered using thermal methods since the thermal resistance of soot particles appears to decrease dramatically with decreasing particle size [27], and/or because of mineral surface-catalyzed oxidation reactions [28]. While the proportion of the soot lost through these mechanisms still remains to be quantified, it is assumed to be proportionally small compared to the soot recovered in the GBC fraction. Non-quantitative recovery of soot upon GBC isolation is however not a critical issue in this work as our main objective is the quantitative density fractionation of soot and graphitic carbon remaining in the mixed GBC fraction.

3.2. *Standards.* The heavy liquid density fractionation method used in this work was validated using aliquots of a 1:1 mixture of a soot sample prepared from n-hexane following the method of Akhter et al. [29], and commercial graphite powder (Table 1). When corrected for the difference in OC concentration of soot and graphite (90.7 and 96.8 wt% OC, respectively), the initial soot:graphite ratio, expressed on a carbon basis, was on average (49.1:50.3) ± 0.53 (n = 5). Since the stable carbon isotope signature of the individual standards was known, the separation efficiency was assessed using two complementary approaches: OC and isotopic mass balances (Table 1).

Table 1. Method validation with *n*-hexane soot and graphite standards

Sample	n	Initial mass[a] (mg)	$\delta^{13}C$ signature (‰)	Recovery based on carbon mass (%)	Recovery based on $\delta^{13}C$ signature (%)
Mixture[b]	5	4.9 - 98.9	-23.17 ± 0.10		
Soot	5	2.5 - 49.3	-23.65 ± 0.07	53.5 ± 2.9	42.3 ± 11.9
Graphite	5	2.4 - 49.6	-22.81 ± 0.18	46.5 ± 2.9	57.7 ± 11.9

[a] The initial masses were 4.9, 12.2, 12.6, 57.0 and 98.9 mg
[b] The soot:graphite mass ratio was about ~1:1 for all experiments

On average, 53.5% of the total OC of the mixture was recovered in the light density fraction (labelled "Soot" in Table 1). No correlation was found between OC recovery and the initial mass of the mixture, suggesting that an initial mass lower than 4.9 mg could be fractionated using this approach. More work is however needed to systematically assess the lower mass limit, as well as the accuracy and precision of the method.

One of the main drawbacks of the OC mass balance assessment of fractionation efficiency is the lack of specificity of the measurement; it is impossible to estimate the purity of the separated fractions by measuring their OC content alone. Fractionation efficiency is better assessed using stable isotopes, provided that the standards are homogenous and isotopically distinct. The $\delta^{13}C$ signatures of the soot and graphite standards used in this work were -23.65 and -22.81 ‰, respectively, giving a difference of 0.84 ‰. Given the instrumental precision of ± 0.1 ‰, the best precision that could be achieved for the isotopic mass balance calculation in this project is roughly ± 12 %, a value that is very close to the precision obtained for the mass balance calculation (± 11.9 %; Table 1). The average recovery of carbon in the low density fraction, calculated from the isotopic signature of the mix and that of the pure standards, was lower than for the OC mass balance calculation (42.3 %). The small difference in the isotopic signature of the two standards could explain why the isotopic mass balance was less accurate and precise than the OC mass balance. Alternatively, incomplete combustion of the more refractory graphite standard could also lead to a lower OC-based assessment of recovery of graphite in the high density fraction; stable isotope-based calculations would not be affected as much by this problem. Adding an oxidation co-catalyst such as vanadium oxide (V_2O_5) to the sample would help eliminate this potential problem. The results from both mass balance approaches however suggest that small amounts of soot and graphite can be fractionated with reasonable precision and accuracy using density fractionation at a density of 1.8 g mL^{-1}, and that OC mass balance calculation should be used when quantifying each component, particularly if the differences in $\delta^{13}C$ between the two fractions are small.

3.3. *Natural samples.* The GBC fraction isolated from a series of natural sediments was fractionated in the same way as for standards (Table 2). OC mass balance calculations were used to quantify the soot and graphitic carbon fractions, and the $\delta^{13}C$ signatures of the separated fractions were measured to probe for differences in the sources and/or dynamics of the two fractions, and also to confirm the OC-based quantitative results.

Soot accounted only for 20.6 to 40.3% of thermally resistant carbon in these samples, confirming the importance of the potential bias when using combustion methods to estimate the soot content of a complex mineral sample (Table 2). As expected, the relative soot contribution was higher in the urban river and lake (Buffalo River and Lake Washington, respectively) than in the Stillaguamish River and the Saguenay Fjord, two sites located away from major sources of direct contamination by soot. OC and stable isotope mass balance calculations generally agreed well, with a larger difference found for the Stillaguamish River sample (13% higher graphitic carbon contribution indicated by the stable isotopes). This result could be explained by the fact that the GBC isolate from this sample consists largely of pure graphite structures, as determined using carbon X-ray absorption near-edge structure spectroscopy and scanning transmission X-ray microscopy [6,7], and that pure crystalline graphite might not be quantitatively combusted in the elemental analyzer.

Surprisingly, the soot contribution to GBC measured in this work for the Lake Washington sample is lower than reported in ref. [21] (41 and 73%, respectively). As the only difference between the two studies is the fact that the samples were combusted in the presence of residual minerals in this work (hard oxide minerals that survive demineralization [15], whereas the low density fraction isolated in [21], was mineral-free), the lower contribution found here could be due to mineral-catalyzed oxidation of the smaller soot particles during combustion, as suggested in ref. [27-28]. Such a catalytic reaction, if occurring, would lead to underestimation of the soot content in mineral-containing samples.

The $\delta^{13}C$ signatures measured for the two fractions suggest a complex picture that cannot be unraveled with the limited number of samples analyzed to date (n = 4). However, two trends emerge. First, soot $\delta^{13}C$ signatures were heavier by about 1.5 ‰ in samples with higher relative soot contributions to GBC (Buffalo River and Lake Washington) compared to samples with a more predominantly graphitic GBC isolate (Stillaguamish River and Saguenay Fjord; Table 2). This result reflects the marine origin (e.g., isotopically heavier) of the soot-producing fossil fuels and suggests a more terrestrial origin, such as forest fires, for soot found in the latter two sites, which are located in densely forested areas. Second, the signature measured for the graphitic carbon fraction appears more depleted by 1-2 ‰ in samples from the eastern part of the continent (Buffalo River and Saguenay Fjord) compared to those from the west coast. With only four samples analyzed, it is however impossible to know whether this is a general trend or just a coincidence.

Table 2. Soot and graphitic carbon in the GBC fraction of natural samples

Sample	n	$\delta^{13}C$ signature (‰)			Contribution to GBC based on OC mass (%)			Contribution to GBC based on $\delta^{13}C$ (%)		
Stillaguamish River sediment										
GBC	2	-24.08	±	0.00						
Soot	2	-25.66	±	0.23	20.6	±	3.5	7.6	±	3.0
Graphitic Carbon	2	-23.95	±	0.09	79.4	±	3.5	92.4	±	3.0
Saguenay Fjord sediment										
GBC	2	-25.71	±	0.06						
Soot	2	-25.58	±	0.25	20.9	±	5.3	25.0	±	32.8
Graphitic Carbon	2	-25.72	±	0.41	79.1	±	5.3	75.0	±	32.8
*Buffalo River sediment**										
GBC	1	-24.70								
Soot	1	-24.08			34.1			47.6		
Graphitic Carbon	1	-25.25			65.9			52.4		
*Lake Washington sediment**										
GBC	1	-23.56								
Soot	1	-24.09			40.3			40.8		
Graphitic Carbon	1	-23.19			59.7			59.2		

* Owing to sample availability, only one $\delta^{13}C$ measurement could be carried out for these samples.

4. Concluding remarks

Heavy liquid fractionation of GBC allows separating two components of presumably different origins, fates, and reactivities. The presence of significant amounts of graphitic black carbon in soot isolates when using chemical or combustion-based methods implies that the global inventories of soot BC should probably be revisited. The lighter fraction (density <1.8 g mL^{-1}), comprising soot-like compounds, most likely belongs to the combustion black carbon continuum, which also includes char and charcoal (Figure 1). Most indications suggest that these components are more labile than generally assumed, particularly with respect to compounds recovered in the high density fraction. This heavier fraction, comprising more crystalline fused polyaromatic materials, appears to be much more refractory and forms, with coal as well as fossil soot, char and charcoal, the slowly-cycling petrogenic black carbon continuum. The life cycle of these high-density components most likely differs considerably from that of the lighter components. Their chemical and isotopic composition, as well as their environmental behavior, including their susceptibility to oxidation, should thus be evaluated separately, which to date can only be accomplished through the method presented here.

References

[1] Hedges, J. I. et al., 2000: The Molecularly-Uncharacterized Component of nonliving organic matter in natural environments. Org. Geochem. **31**, 945-958.

[2] Komada, T., Schofield, O. M., Reimers, C. E., 2002: Fluorescence characteristics of organic matter released from coastal sediments during resuspension. Mar. Chem. **79**, 81-97.

[3] Nelson, Baldock, J. A., 2005: Estimating the molecular composition of a diverse range of natural organic materials from solid-state ^{13}C NMR and elemental analyses. Biogeochem. **72**, 1-34.

[4] Hsu, P.-H., Hatcher, P., 2006: Covalent coupling of peptides to humic acids: Structural effects investigated using 2D NMR spectroscopy. Org. Geochem. **37**, 1694-1704.

[5] Hertkorn, N., et al., 2006: Characterization of a major refractory component of marine dissolved organic matter. Geochim. Cosmochim. Acta **70**, 2990-3010.

[6] Haberstroh, P. R., Brandes, J. A., Gélinas, Y., Dickens, A. F., Wirick, S., Cody, G. D., 2006: Chemical composition of the graphitic black carbon fraction in riverine and marine sediments at sub-micron scales using carbon X-ray spectromicroscopy. Geochim. Cosmochim. Acta **70**, 1483 - 1494.

[7] Brandes, J., Rumble, D., Haberstroh, P., Wirick, S., Gélinas, Y., Cody, G.D., 2008: Carbon XANES spectroscopy of natural graphite. Carbon, doi:10.1016/j.carbon.2008.06.020.

[8] Olcott, A. N., Eglinton, T. I., 2008: Chemical and isotopic characterization of products from ruthenium tetroxide oxidation of macromolecular organic matter in marine sediments. ASLO Meeting, March 2-7, Orlando (FL), USA.

[9] Eusterhues, K, Rumpel, C., Kleber, M., Kögel-Knabner, I., 2003: Stabilization of soil organic matter by interactions with minerals as revealed by mineral dissolution and oxidative degradation. Org. Geochem. **34**, 1591-1600.

[10] Masiello, C. A., 2004: New directions in black carbon organic geochemistry. Mar. Chem. **92**, 201–213.

[11] Knicker H., Hilscher A., González-Vila F.J. and Almendros G. 2008: A new conceptual model for the structural properties of char produced during vegetation fires. Org. Geochem. **39**, 935-939.

[12] Kuo L.-J., Herbert B.E. and Louchouarn P. 2008: Can levoglucosan be used to characterize and quantify char/charcoal black carbon in environmental media? Org. Geochem. In press.

[13] Gustafsson, O., Haghseta, F., Chan, C., Macfarlane J., Gschwend, P.M., 1997: Quantification of the dilute sedimentary soot phase: implication for PAH speciation and bioavailability. Environ. Sci. Technol. **31** 203–209.

[14] Cachier H., Brémond M. P. and Buat-Ménard P. 1989: Determination of atmospheric soot carbon with a simple thermal method, Tellus, **41B**, 379-390.

[15] Gélinas, Y., Prentice, K. M., Hedges, J. I., Baldock, J. A., 2001: An improved thermal oxidation method for the quantification of soot/graphitic black carbon in sediments and soils. Environ. Sci. Technol. **35**, 3519 - 3525.

[16] Gustafsson, O., et al., 2001: Evaluation of a protocol for the quantification of black carbon in sediments. Global Geochem. Cycles **15**, 881-890.

[17] Maillard, L. C., 1912: Comptes-Rendus de l'Académie des Sciences (Paris) **156**, 148-149.

[18] Louchouarn P. *et al.* (2007). Elemental and isotopic evidence of soot- and char-derived black carbon inputs to New York City's atmosphere during the 20th Century. Environ. Sci. Technol. **41**, 82-87.

[19] Hammes, K., et al., 2007: Comparison of black carbon quantification methods using reference materials from soil, water, sediment and the atmosphere, and implications for the global carbon cycle. Global Biogeochem. Cycles **21**, 10.1029/2006BG002914.

[20] Dickens, A. F., Gélinas, Y., Masiello, C. A., Wakeham, S. G., Hedges, J. I., 2004: Reburial of fossil organic carbon in marine sediments. Nature **427**, 336 - 339.

[21] Dickens, A. F., Gélinas, Y., Hedges, J. I., 2004: Physical separation of combustion and rock sources of graphitic black carbon in sediments. Mar. Chem. **92**, 215 - 223.

[22] Stanmore, B. R., Brilhac, J. F., Gilot, P., 2001: The oxidation of soot: a review of experiments, mechanisms and models. Carbon **39**, 2247-2268.
[23] Wakeham, S. G., Masiello, C. A., Gélinas, Y., Alexander, C. R., Leavitt, P., 2004: Hydrocarbon in Lake Washington Sediments - A 25-year retrospective. Environ. Sci. Technol. **38**, 431 - 439.
[24] Kim, W.-S., Park, Y. H., Shin, J. Y., Lee, D. W., Lee, S., 1999: Size determination of diesel soot articles using flow and sedimentation field-flow fractionation. Anal. Chem. **71**, 3265 - 3272.
[25] Choi, M. Y., Mulholland, G. W., Hamins, A., Kashiwagi, T., 1995: Comparisons of the soot volume fraction using gravimetric and light extinction techniques. Combust. Flame **102**, 161-169.
[26] Pierson, H. O., 1993: The handbook of carbon, graphite, diamond and fullerene: properties, processing and applications. Noyes Publications, Park Ridge.
[27] Lima, A. L., 2004: Molecular and isotopic records of combustion inputs to the environment over the last 250 years. Ph.D. Thesis, Woods Hole Oceanography Institution. 267 p.
[28] Elmquist, M., Gustafsson, O., Andersson, P., 2004: Quantification of sedimentary black carbon using the chemothermal oxidation method: an evaluation of ex situ pretreatments and standard additions approaches. Limnol. Oceanogr. Methods **2**, 417-427.
[29] Akhter, M. S., Chughtai, A. R., Smith, D. M., 1985: The structure of hexane soot I: Spectroscopic studies. Appl. Spectros. **39**, 143 - 153.

IOP Conf. Series: Earth and Environmental Science **5** (2009) 012011 doi:10.1088/1755-1307/5/1/012011

Is it possible to use "twin cores" as a unique sedimentary record? *An experimental design based on sediment color*

C. Veiga-Pires[1], **N. C. Mestre**[2]

1 Universidade do Algarve, CIMA Research Centre, Campus de Gambelas, 8005-139 Faro, Portugal

2 School of Ocean and Earth Science, National Oceanography Centre, Southampton, University of Southampton, European Way, Southampton, S014 3ZH, UK

E-mail: cvpires@ualg.pt

Abstract. Sedimentary cores are widely used for studying Quaternary records. However, the amount of sediment that is available is proportional to the diameter of the core, which is rarely bigger than 15 cm. One way to obtain more sediment is to use two cores retrieved from almost the same location and use them as if they represent a unique sedimentary record. In the present work, an experimental design has been applied to verify if "twin cores" from an estuary can be considered as representing the same sedimentary record with twice the amount of sediment to study. Because sediment can be characterized based on its color, the variables used as replicates in the experimental design are the three Lab CIE colors acquired with a X-Rite Colortron spectrophotometer. Sediment cores were retrieved from the upper saltmarsh of Gilão River's estuary, southern Portugal. Twin cores, with in between distances of 50 cm, 100 cm and 200 cm, from two different sites were analysed. Results from a nested ANOVA show that even for the closest twin cores (50 cm apart) there is at least one color variable that shows significant variations between the profiles of both cores. These results clearly show that "twin cores" cannot be used as a unique sedimentary record without any previous testing, at least in such transitional regions.

1. Introduction

The study of sedimentary cores allows the reconstruction of the environmental history of marine e.g. [1], transitional e.g. [2] and continental regions e.g. [3]. Most of the coring apparatus available have restricted diameter dimensions (3 to max 20 cm) [4], limiting the quantity of sediment representative of one deposition layer or age interval. In such conditions, it is very difficult to perform all the required or suitable analyses as most of them imply the destruction of the scarce amount of sediment available per time interval. Therefore, there is the temptation to use a "twin core" in order to get more sediment per time interval and complete the series of analyses on the "equivalent" sediment record. This is mostly true in study areas such as transitional zones, namely estuaries, where it is easy to get the "twin cores" done almost exactly in the same place with no additional cost. In this context, the present work aims to verify the possibility of using such "twin cores" for analytical purposes as if they were a unique sedimentary record.

In order to answer this question, an experimental design has been set up based on the color of the sediment because it is one of the sediment characteristics considered to represent an integrated signal of the sediment composition and commonly used for stratigraphic or correlation purposes [5]. Moreover, the digital color, which can be acquired with a spectrophotometer through non destructive manipulation, gives quantitative information on the color and is used to describe the sediment as well as to document changes in depositional conditions [6]. It is also important to refer that the acquisition of digital color is an affordable and non-destructive analysis that allows further investigation on the sediment afterwards. On such basis, the digital color of the sediment is the parameter considered in the present study as the indicator for similarity between the "twin cores".

2. Material and methods

In order to determine the possibility of replication of cores for sampling and sub-sampling, an experimental design was elaborated previously to the sampling of the sedimentary cores. By definition, the experimental design is developed to outline the way in which the experiment, namely the sampling, should be carried out in order to obtain data that will answer the scientific question by means of statistical values.

2.1. Experimental design

The present work aims to determine whether "twin cores" (or paired cores) can be treated as duplicate records to increase the sample volume collected in the field. To answer this question it is important to obtain statistically significant results, meaning that the effect of the distance between cores has to be evaluated statistically, that the observed pattern is not due to random or to site specific conditions.
In order to address all these questions, the experiment has been designed considering the following null hypothesis: "The color is the same for each depth of each core from one site no matter the distance between the cores" in order to find out how far can two cores be collected and be used as "twin cores". If the null hypothesis is true, (i.e. p-value not significant) then "twin cores" can be used as a unique record, if null hypothesis is not true (i.e. p-value significant) then alternative hypothesis is verified. The alternative hypothesis is "The color profiles are different between cores ".

The experimental design based on the above questions contains several levels (factors): 1) the site sampling, 2) the distances between "twin cores" that are reproduced in each site, 3) the two cores sampled for each distance at each site, 4) the depth sampling in each core, and 5) the replicates of the data acquisition at each depth (figure 1, table 1).

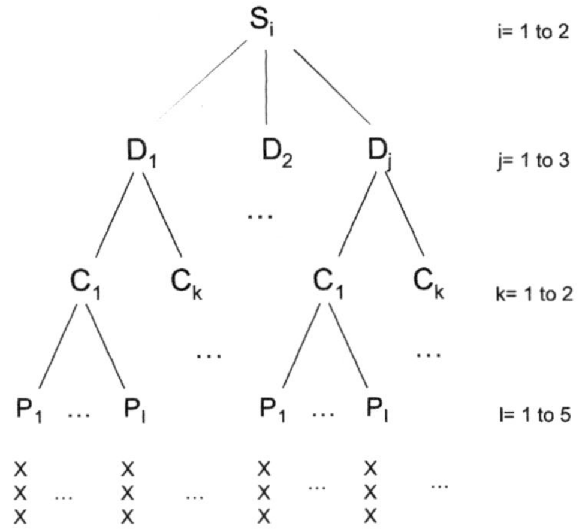

Figure 1. The overall design consists in having two different sampling sites (Si) in order to verify that the results are not just due to a specific site; three different distances (Dj) between the twin cores (Ck); depths (Pl) in each core, which can be considered independent if separated by at least 20 cm; and finally three replicates of colour at each depth (x).

This experimental design is therefore considered as hierarchical, or nested, and deals with the interaction that could exist between the distances between cores and the sampling site. In other words, this experiment should allow comparing the variations between sites as well as within site, and

IOP Conf. Series: Earth and Environmental Science **5** (2009) 012011 doi:10.1088/1755-1307/5/1/012011

between cores as well as within cores, with a statistical value for significance. The number of variables for each factor, and therefore the degrees of freedom, have been set in order to optimize the statistics [7] as describe in table 1, for a total degree of freedom of 179.

Table 1: Components of the planned experimental design

Factors	Source of variation	Number	Type	Degree of freedom
Site	S_i	2	Random	1
Distance between cores	D_j	3	Fixed Orthogonal	2
	S_i x D_j		Interaction	2
Cores	$C(S,D)_{k(i,j)}$	2	Random nested	6
Depths	$P(C(S,D))_{l(k(i,j))}$	5	Random nested	48
Replicates	$E_{m(l(k(i,j)))}$	3	Residuals	120
Total				**179**

2.2. Sampling

The sampling site chosen for the present study is a saltmarsh area (37° 11' N, 07° 27' W) located in the Gilão River estuary, near Tavira's town, South Portugal (figure 2). Sampling was carried out with a half-cane hand corer (figure 3) allowing the retrieval of sediment records of up to 3.15 meters deep, in sections of 1 m long and 3 cm in diameter, from the upper marsh of the Gilão River estuary (figure 4). Noteworthy is the fact that sampling occurred during summer, sampling sites being very dry as can be observed in figure 4. The recovered sediments were mainly constituted by very fine to medium silt, with almost homogeneous granulometry along the core, but generally two different color sections from top to bottom. Due to highly dense root development, the top sediment was rarely recovered. Moreover, none of the cores reached the water table level implying that no drastic moisture change along the cores was observed. Sediment was then transferred to rectangular tubes used in electrical assembly in order to be transported in the same day to the laboratory. On arrival in the laboratory, the three color replicates were immediately acquired for each core using the spectrophotometer X-Rite Colortron™. The success of this sampling phase was decisive to confirm the statistical model to be used for the analysis of variance (section 2.3).

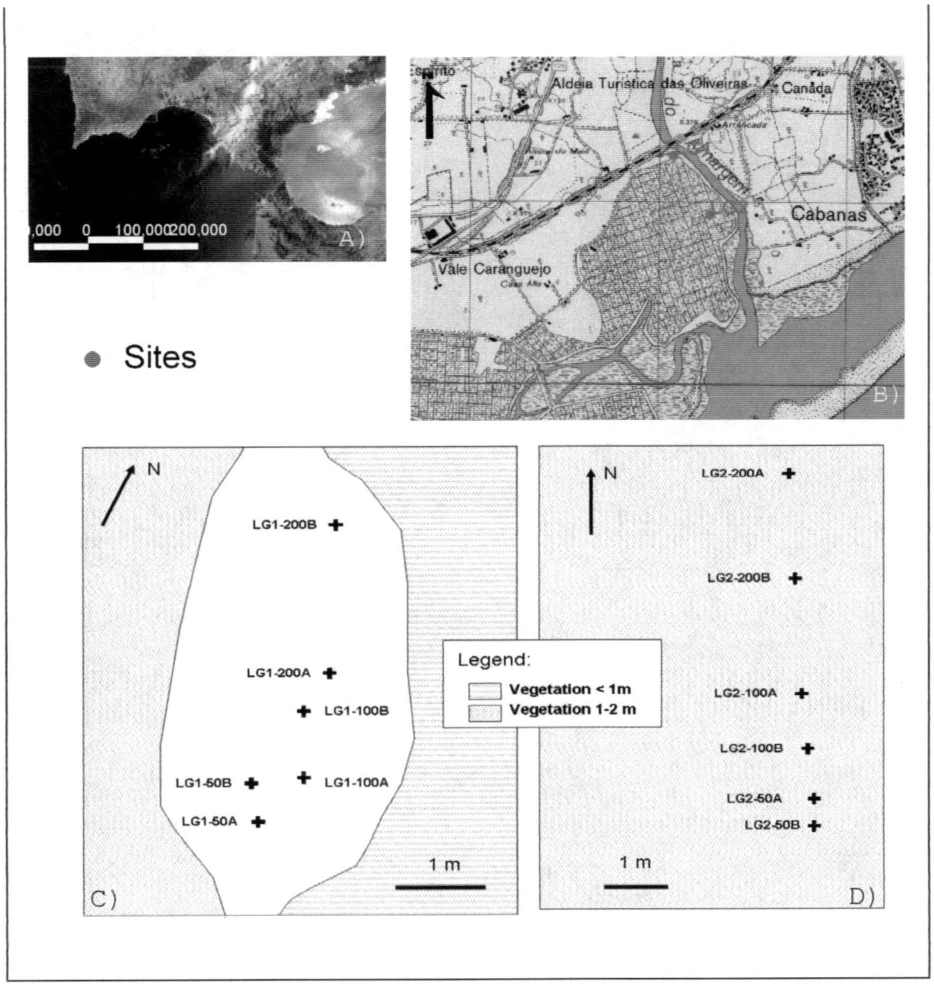

Figure 2. A) Satellite view of the Iberian Peninsula with the location of the area of the Almargem-Gilão river estuary; B) Close up of the region topographic map with the location of the two sites used in the present study that are ~400 m apart: LG1 and LG2; C) and D) Oriented schemes of the two sites and the cores retrieved in each one of them, for site LG1 and LG2 respectively.

Accordingly, several methodological tests have been realized, prior to the present work, for optimizing the color data acquisition and its reproducibility. The resulting protocol consist in covering the surface of the core with transparent 5 cm large scotch tape to insure that there is no contamination by the spectrophotometer and that the sediment is not oxidized. The spectrophotometer X-Rite Colortron™ was calibrated prior to analyzing each data series, using a 100 % reflectance pure white, followed by a 0 % reflectance black box, and finally by 6 colors' target [8]. The 3D color space chosen for the present work was the *Commission Internationale de l'Éclairage* (CIE) Lab , where **L** values represent the brightness (0 to 300), **a** values represent the color on the red-green axis (-300 to +300) and **b** values represent the color on the yellow-blue axis (-300 to + 300) [8]. Each of the color components Lab are treated separately in the experimental design statistics.

IOP Conf. Series: Earth and Environmental Science **5** (2009) 012011 doi:10.1088/1755-1307/5/1/012011

Figure 3. Half-cane hand corer apparatus which has a diameter of ca 3 cm.

Figure 4. Picture showing the remained holes after collecting the sedimentary cores at one of the two sites (LG1). Note that the second site was chosen to have similar environmental conditions, for instance vegetation.

2.3. Statistics

The statistical model for the analysis of variance (ANOVA) of the present experimental design is a nested linear model that can be expressed as follows:

$$X_{ijklm} = \mu + S_i + D_j + C(S,D)_{k(i,j)} + P(C(S,D))_{l(k(i,j))} + E_{m(l(k(i,j)))} \quad \text{(cf table 1 for details)} \quad (1)$$
$$H_0: \mu = \mu_1 = \ldots = \mu_n$$
$$H_A: \mu \neq \mu_1 \neq \ldots \neq \mu_n$$

Where X_{ijklm} is the mean of values (namely CIE L values, or CIE a values, or CIE b values), μ is the variance, S_i is the variation effect related to sites (i= 1 to 2), D_j is the variation effect related to distance (j= 1 to 3), $C(S,D)_{k(i,j)}$ is the variation effect related to cores (k= 1 to 2), $P(C(S,D))_{l(k(i,j))}$ is the variation effect related to depth (l= 1 to 5) and $E_{m(l(k(i,j)))}$ are the residuals due to replication (m = 3).

The null hypothesis (H_0) implies that the color of the sediment (and therefore its composition) is the same for each depth of each core no matter the distance between cores and/or the site location. The alternative hypothesis (H_a) implies that there are differences in the color between depths, which is expected, and/or between cores at different distances. If the latter is verified for any of the three distances between cores, then there are no so called “twin cores”.

Statistics have been computed under R program [9] using the linear model of equation 2.1 in “aov” function[1] [10]. Since the cores did not have the same recuperation length and color data were acquired every 5 cm along the depth (figure 5), color data had to be filtered in order to obtain a homogeneous matrix, *i.e.*, with no row with empty cell [11]. In consequence, only five common depths of color data were comparable for all the cores. The resultant matrix is built on 5 sediment depths that are 20 cm

[1] aov (Color ~ factor(Site) + factor(Distance) + factor(Distance)*factor(Site) + factor(Site):factor(Distance):factor(Core) + factor(Site):factor(Distance):factor(Core):factor(Depth))

apart, for which independency have been verified through the lack of autocorrelation even if the sample number is reduced (not shown here), starting at 1.55 m and ending at 2.35 m deep, forming a balanced dataset with no missing cells [12].

3. Results and discussion

The results of the analysis of variance for the three color parameters of the color space CIE Lab are expressed in table 2. Significant statistical differences were found for all sources of variation in the model with *p*-values <0.001 (***), except for the interaction between site and distance for the reflectance (CIE L) with a *p*-value =0.001 (**; table 2). Not shown here is the distribution of residuals and standardized residuals that show no heterogeneity [13]. Therefore, the results imply that the null hypothesis can be rejected and that there is a difference in the three color components between the cores taken at distances of 200 cm, 100 cm or even 50 cm apart. The fact that there are significant differences between depths (figure 5) is coherent since those differences are used to document the sediment history [14] . It was also expected to obtain significant differences between sites since they are 400 m apart and an estuarine environment is very complex with a variety of different depositional features such as channels and sand banks that change laterally over decimetric scale [15].

The differences in the depth profiles of each color parameter between cores at both sites (figure 5) clearly illustrate the statistical results (table 2). Note that depth profiles between cores A & B do not show any significant mismatch (cross-correlation significant at lag= 0, not shown here). The reflectance profiles (CIE L) show almost no overlap for both distances between cores, 50 cm (figure 5A) and 200 cm (figure 5B), at both sites (LG1 and LG2). Although, it is also interesting to note that the deepest part of the CIE L profiles of the two cores, 50 cm apart, from site LG2 (figure 5A) fits perfectly and could partially be interpreted as a "twin record". This observation is important because it could open the possibility to find "twin records" as punctual and exceptional cases but not as a rule. The CIE L variable, also known as reflectance, is believed to be linked either to the water content of the sediment [16], its grain size [17], or even its carbonate or silica content [18].

Table 2: Results from the analysis of variance based on the linear model from equation 1 for each of the three color parameters. Df : Degree of freedom, Sum Sq.: Sum squared, Mean Sq.: Mean squared

Source	Df	Sum Sq.	Mean Sq	F value	P value	Signif.
Site	1	2863.78	2863.78	1749.5036	< 2.2e-16	***
		214.71	214.71	1500.413	< 2.2e-16	***
		792.08	792.08	926.0158	< 2.2e-16	***
Distance	2	159.62	79.81	48.7575	3.173e-16	***
		14.08	7.04	49.205	2.481e-16	***
		14.45	7.23	8.4489	0.0003690	***
Site*Distance	2	23.03	11.52	7.0349	0.001291	**
		13.74	6.87	48.023	4.766e-16	***
		42.59	21.29	24.8952	9.038e-10	***
Core	6	1235.54	205.92	125.7994	< 2.2e-16	***
		15.53	2.59	18.087	7.262e-15	***
		254.75	42.46	49.6370	< 2.2e-16	***
Depth	48	2851.06	59.40	36.2860	< 2.2e-16	***
		521.91	10.87	75.983	< 2.2e-16	***
		1919.57	39.99	46.7534	< 2.2e-16	***
Residuals	120	196.43	1.64			
		17.17	0.14			
		102.64	0.86			

Signif. codes: 0 '***' 0.001 '**' 0.01 '*' 0.05 '.' 0.1 ' ' 1
Black (CIE L); Blue (CIE a); Green (CIE b)

In the present case, the retrieved sediment was relatively dry (figure 4) but no water content was measured enabling to put aside the assumption than the observed differences in variance could be linked to changes in sediment moisture and not in composition. Indeed, a previous work on 12 cores from the upper marsh of 4 estuaries from the Algarve region showed that the moisture of the sediments do change significantly the CIE L values (p-value=0.2674 with two-sample t-test on n=144) but not the other two color components (figure 6).

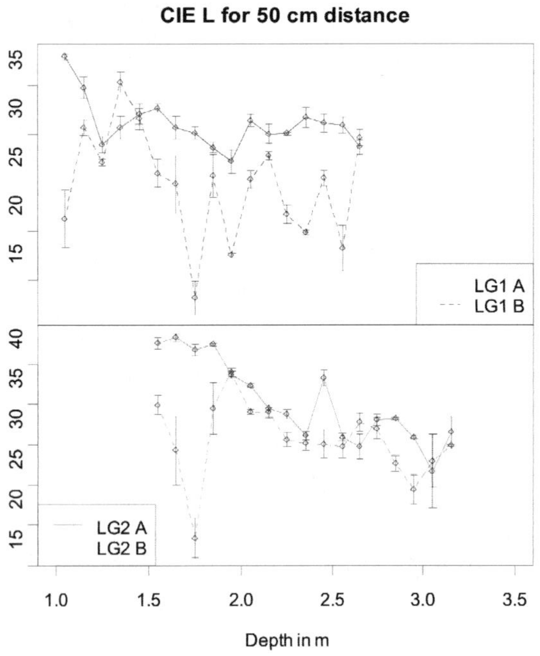

A

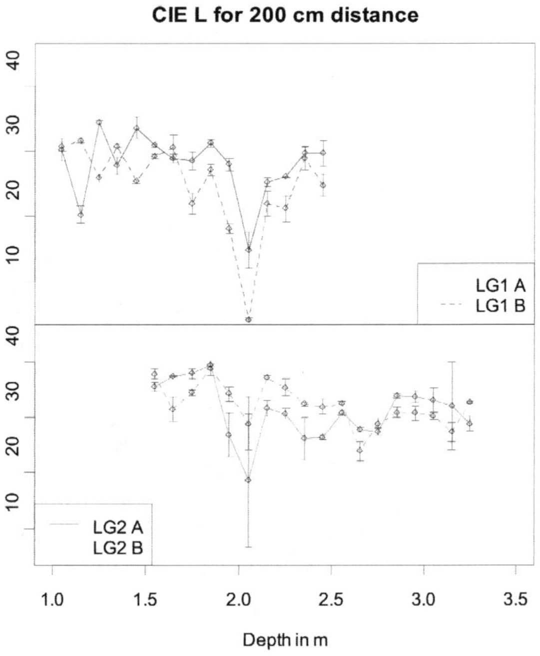

B

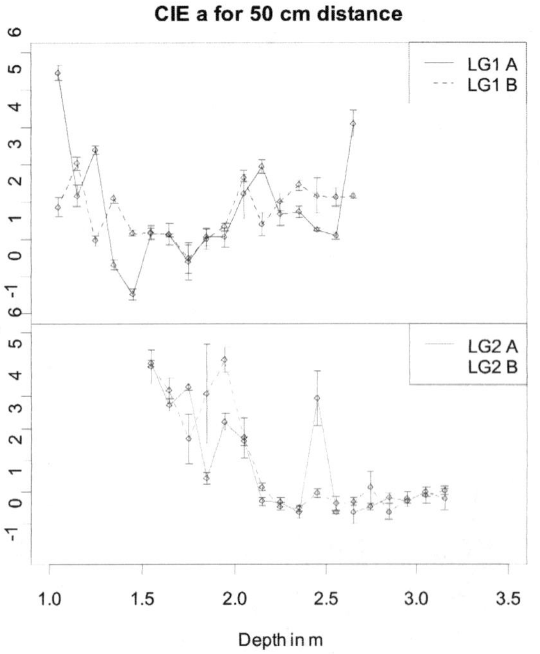

C

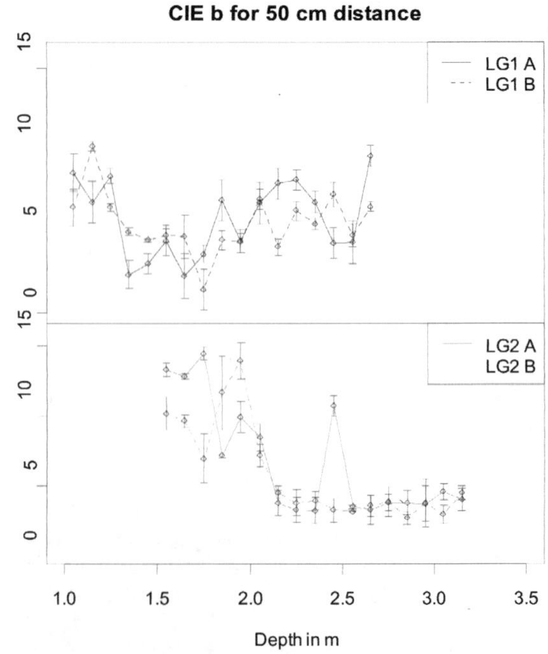

D

Figure 5. Depth profiles of the mean values and the standard deviations, at both sites LG1 and LG2, for: A) CIE L for cores distant by 50 cm; B) CIE L for cores distant by 200 cm; C) CIE a for cores distant of 50 cm; and D) CIE b for cores distant of 50 cm (see text for discussion)

Hence, as the other color parameters profiles (CIE a and CIE b; figure 5C and 5D) show no corresponding overlap, one should not regard two records as similar or "Twin Cores" based only on the CIE L parameter profile. Likewise, the cores LG1A and LG1B show apparent overlap in the shallowest part of the profile of the CIE a and b parameters (figure 5C and 5D), but it is not statistically significant.

In general the color profiles (figure 5) reflect that: i) there are significant differences between the two sites; ii) there are significant differences between the two cores at any site and at any distances between them and iii) there are differences within depths. With increasing depth, the difference in the mean values of the color parameters between cores tends to be smaller, as well as the standard deviations. On the other hand, the upper part of the profiles shows great variability in mean values between cores with high standard deviations (figure 5). Assuming that the sediment accumulation rates are similar, this might be related to the upper sedimentary "instability" linked to vegetation roots, redox front and dynamics of groundwater table, which characterize saltmarsh environments [19]. At greater depth, the chemical conditions are more stable and thus the variance observed in sediment color between cores and even between sites should be smaller.

When plotting the main effects of the four variables used in the linear model on the mean values of each color parameter (figure 7), it appears that the variance for the deeper samples is effectively smaller than the variance for the shallower ones. In addition, the three color variables have different values for the upper part of the cores in relation to the deeper part. This corroborates the above observed variability in color profiles.

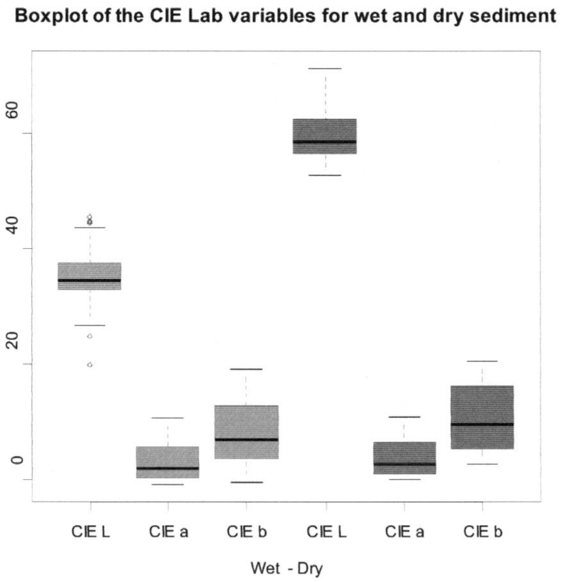

Figure 6. Boxplot of the three CIE lab color components for wet (green) and dry (red) sediment. Each boxplot represents 144 color analyses on samples recovered from12 cores from the Algarve region. Two-sample t-test between the wet and dry CIE L data sets gives a p-value of 0.2674 which means that the two data sets are significantly different.

Furthermore, figure 7 puts in evidence that the main effects on the means' variance are due to site location and sample depth but with much less contribution from the distance between cores. However, it is interesting to note that main variations in CIE a values seem to be linked to depth and not sites which can be interpreted as an indicator of a reduce lateral variation for sediment characteristics that are registered through this color parameter. Therefore, a detailed description of all interactions and

IOP Conf. Series: Earth and Environmental Science **5** (2009) 012011 doi:10.1088/1755-1307/5/1/012011

effects for each color parameter and at each factor level, as presented in the experiment design (table 2 and figure 7), can provide important information, as long as the link between color and sediment characteristics is known. In any case, if the purpose of a study is to compare records between sites, twin cores could still be used because the effect of distances is much less important than the one of sites.

A

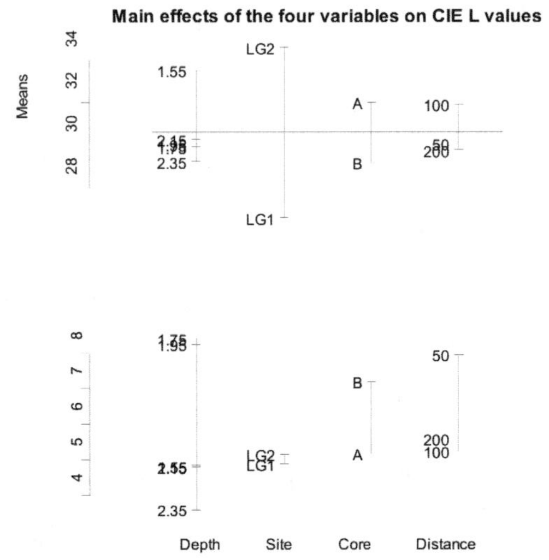

B

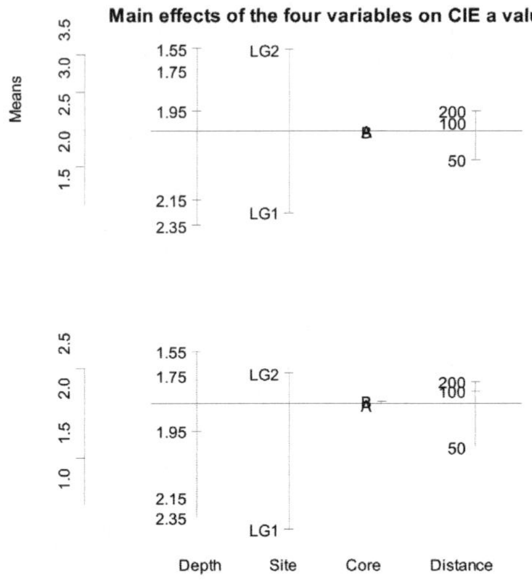

C

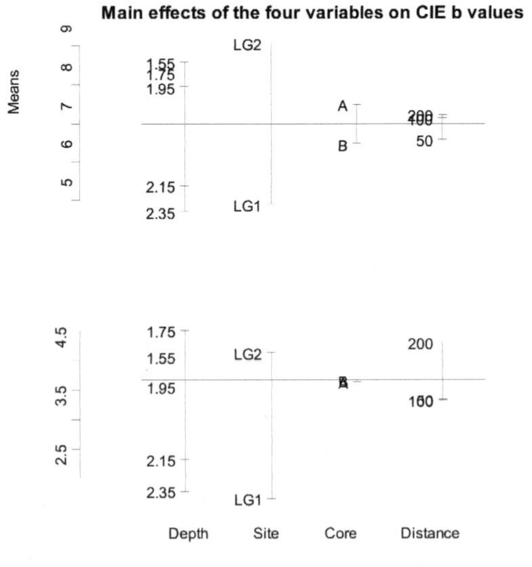

Figure 7. Representation of the main effects of the four variables used in the experimental design on the means and standard deviations of the values obtained for each colour axis: CIE L (A), CIE a (B) and CIE b (C). Horizontal line in each of the design plot represent the overall mean of the data set (n=180) For instance for the mean of all CIE L values is 30.61 , 2.49 for all CIE a values, and 6.98 for all CIE b values.

The present study used sample depth intervals of 20 cm in order to have independence between samples but the general trend in cored sediment's studies is to work at very high resolution in order to acquire (almost) continuous physical properties [20] for documenting short time climatic events [6]. Such high resolution studies when applied to depositional environments such as estuaries where lateral variations are very important should have strong concern in extracting the overall climatic record from a unique site [2], because the variance observed between sites of only some hundreds of meters away might be more important, in some cases, than the variance within the site or the core.

In conclusion, the present experimental design helps to document the possibility of sediment core duplication for being used as “twin cores” for analysis purposes whenever sampling techniques allows it. This field of statistics, among others, allows including confidence and levels of significance to relations that might be either biased by the observer background or by the sampling method. Indeed, in the present study, the structure of the experimental design permits to conclude that the results are not due mere coincidence or randomness, at least for the studied estuary. A similar experimental design should be applied in future studies aiming to collect twin cores in such transitional regions.

Acknowledgment
The authors acknowledge the two reviewers (anonymous and M. Debret) for their constructive comments. This work was part of a financed project by FEDER and OE through the Portuguese Foundation for Science and Technology (FCT) (POCTI/CTA/39733/2001).

References

[1] Andrews J T and Freeman W 1996 The Measurement of Sediment Color Using the Colortron™ Spectrophotometre *Arc. Alp. Res.* **28** 524-528
[2] Boski T, Camacho S, Moura D, Fletcher W, Wilamowski A, Veiga-Pires C, Correia V, Loureiro C and Santana P 2008 Chronology of the sedimentary processes during the postglacial sea level rise in two estuaries of the Algarve coast, Southern Portugal *Estuar. Coast. Shelf Sci.* **77** 230-244
[3] Stuiver M 1970 Oxygen and carbon isotope ratios of fresh-water carbonates as climatic indicators *J. Geophys. Res.* **75** 5247–5257
[4] Rothwell R G and Rack F R 2006 *New Techniques in Sediment Core Analysis* ed R G Rothwell *(Geo. Soc. Spec. Pub..* **267***)* (London:Geological Society) pp1-29
[5] Mix A C, Rugh W Pisias N G, Veirs S, Leg 138 Shipboard Sedimentologists and Leg 138 Scientific party 1992 *Color Reflectance Spectroscopy: A Tool for Rapid Characterisation of deep-sea sediments* (*Proc. of ODP* **138**) ed N G Pisias *et al* (Texas: Ocean Drilling Program) pp 67-77
[6] Nederbragt A Jand Thurow J W 2004 *Digital Sediment Colour Analysis as a Method to Obtain High Resolution Climate Proxy Records* ed P Francus (Dordrecht: Springer) pp 105-124
[7] Underwood A J 2006 *Experiments in Ecology: Their Logical Design and Interpretation using Analysis of Variance* (Cambridge: Cambridge University Press) p 504
[8] X-Rite Color 2000 *ColorShop – Getting Started Guide* p 70
[9] R Development Core Team 2007 R: A language and environment for statistical computing R Foundation for Statistical Computing (Vienna) http://www.R-project.org
[10] Maindonald J and Braun J 2007 *Data analysis and Graphics using R an Example-Based Approach 2nd Edition* ed R Gill et al (New York: Cambridge University Press) p 502
[11] Dalgaard P. 2002 *Introductory Statistics with R* ed J Chambers Use *et al* (New York: Springer) p 267
[12] Quinn G P and Keough M J 2002 *Experimental Design and Data Analysis for Biologists* (Cambridge: Cambridge University Press) p 537
[13] Zuur A F, Ieno E N and Smith G M 2007 *Analysing Ecological Data* ed M Gail et al (New York: Springer) p 672
[14] Francus P, Bradley R S and Thurow J W 2004 *An Introduction to Image analysis, Sediments and Paleoenvironments* ed P Francus (Dordrecht: Springer) pp 1-7
[15] Walker R G and James N P 1992 Facies Models: Response to Sea Level Change (Stistville, Ontario: Geological Association of Canada) p 409
[16] Nagao S and Nakashima S 1991 A convenient method of color measurement of marine sediments by colorimetry *Geochem. J.* **25** 187-197
[17] Blum P 1997 Physical Properties handbook: a guide to the shipboard measurement of physical properties of deep-sea cores *ODP Tech. Note* ***26***

[18] Balsam W L and Deaton B C 1991 Sediment dispersal in the Atlantic Ocean – Evaluation by visible light spectra *Rev. in Aquatic Sci.* **4** 411-447

[19] Fagherazzi S, MArani M and Blum L K 2004 Coastal Estuarine Studies: The Ecogeomorphology of Tidal Marshes (Washington: AGU) p 268

[20] St-Onge G, Mulder T, Francus P and Long B 2007 *Continuous Physical Properties of Cored Marine Sediments* ed C Hillaire-Marcel and A de Vernal (Oxford: Elsevier) pp 63-98

IOP Conf. Series: Earth and Environmental Science **5** (2009) 012012 doi:10.1088/1755-1307/5/1/012012

The use of fish remains in sediments for the reconstruction of paleoproductivity

T Drago[1, 3], V Ferreira-Bartrina[2], A M P Santos[1] and J Pinheiro[1]

[1] Instituto Nacional de Recursos Biológicos (INRB), L-IPIMAR, Av. 5 de Outubro s/n 8700-305 OLHÃO, Portugal

[2] Centro de Investigación Científica y de Educación Superior de Ensenada- CICESE, Km.107 Carretera Tijuana, C.P.22860, Ensenada, B.C., Mexico

E-mail: tdrago@ipimar.pt

Abstract. The majority of the works concerning fish productivity are based in fish landing records. However, in order to understand the causes of variability in fish productivity (natural and/or anthropogenic) it is essential to have information from periods when human impacts (e.g., fisheries) are considered unimportant. This can be achieved through the use of fish remains, i.e. scales, vertebrae and otoliths, from sediment records. The obtained data can be used to develop time series of fish stocks revealing the history of fish population dynamics over the last centuries or millennia. The majority of these works are located in Eastern Boundary Current Systems (e.g., Benguela, Peru-Humboldt, California), because these are associated with coastal upwelling and high productivity, which in some cases is at the origin of low bottom oxygen levels, leading to scale preservation. A search for fish remains in the Portuguese margin sediments is in progress in the context of the ongoing research project POPEI (High-resolution oceanic paleoproductivity and environmental changes; correlation with fish populations), which intend to fill the gap in studies of this type for the Canary Current System. In this paper we review some general ideas of the use of fish remains, related studies, methodologies and data processing, as well as presenting the first results of POPEI.

1. Introduction

Changing conditions in oceanic productivity have been under close scrutiny during the last decades. These fluctuations generate important consequences on human societies dependent on marine resources. To understand the role and importance of natural fluctuations and the true extent of human impact on marine productivity we need data from periods where anthropic action was negligible. A substantial amount of data describing oceanic productivity and its fluctuations acquired since the beginning of the instrumental period is currently available, but paleoproductivity data of comparable resolution is scarce or nil.

[3] To whom any correspondence should be addressed.

Several studies based on fisheries records from the last hundred years indicate that important decadal changes have occurred in the productivity of different fish species all around the world [1, 2, 3]. Fluctuations in fish productivity on the Canary Current/Iberian Peninsula Upwelling Ecosystem have also been registered [4, 5]. These fluctuations seem to be largely environmentally driven, mainly related to upwelling strength variability [6, 4, 5]. However, because these studies are based mainly in commercial fish catches, it is difficult to understand the relative importance between natural and/or human-induced forcing. In that sense, sedimentary records of fish stocks could provide indications of such fluctuations in the absence of industrial fishing records, and combined with other biological productivity proxies from the same sedimentary record, help support inferences about the relationship between environmental processes and fish productivity.

The use of fish scales and other fish remains such as bones (vertebrae, spines) and otoliths from marine sediments have been used to construct time series of fish stocks revealing the history of fish population dynamics over the last thousand years. After the pioneer work of Soutar in the late 1960s in Santa Barbara Basin, California [7, 8, 9], subsequent works focused in the Southern Californian coast [10], Central Gulf of California, Mexico [11], slope sediments off Peru in the Humboldt Current system [12, 13], on some fjords, such as the Saanich and Effingham inlets in British Columbia, Canada [14, 15] and off Southwest Africa margins in the Benguela Current system [16]. The majority of these works are located on Eastern Boundary Current Systems (Benguela, Peru-Humboldt, California). These places are, per excellence, associated to regions with coastal upwelling, having high biological productivity, and supporting dense concentrations of small pelagic fishes, and where the low oxygen concentration of the overlaying waters permits the occurrence of primary sedimentary structures, formed at yearly frequencies and preserved because of the lack of bioturbation from benthic fauna.

Until now, no similar work has been done in the Canary Current Upwelling Ecosystem. POPEI- "High-resolution oceanic paleoproductivity and environmental changes; correlation with fish populations" a research project funded by the Portuguese Science Foundation (FCT) and coordinated by INRB-IPIMAR, has the objective of filling part of this gap by studying the Portuguese continental margin. However, the Portuguese margin is exposed to the Atlantic Ocean wave climate, and is consequently characterized by an higher energetic oceanographic environment which leads to higher oxygenation of the sea bottom, thus presenting very different conditions from the basins and fjords where the existing studies were done.

Besides this attempt to find fish remains in sediments, in order to understand the present and past fluctuations of fish populations (eventually before major anthropic impacts), the POPEI project also aims for a better perception of productivity oscillations in general, based on other productivity proxies as calcareous nanoplankton, foraminifera and diatoms. The correlation between fish remains and other proxy data not only will help to correlate fish population's fluctuations with oceanic productivity changes but also will be important to infer the respective forcing factors. We hope that these data will be useful to produce relevant information for management fisheries policies.

In the frame of this project, a regional photographic atlas of fish scales and bones for the common species of the region is being developed 'figure 2'. In January 2008, during the first sea mission, grab samples, gravity cores and a 5 m long vibrocore were collected. Some scales and vertebrae were already found 'figure 4' and cores are being documented by x-ray images 'figure 1'.

This paper aims to review the essential concepts about fish remains found in the sedimentary record and their possible application as indicators of past fluctuations of fish populations, in particular and ocean productivity, in general, and to present the ongoing activities and first preliminary results of the work conducted off the Iberian Upwelling System under the POPEI project.

2. General information about fish remains

Fish remains found in sediments include scales, bones and otoliths. The first ones are the most common, and also the most easily identifiable. Scales are composed by a mixture of micro-crystalline apatite and hydroxy apatite, embedded in a collagen-icthylepidin matrix [17]. As the fish grow, scales grow laterally through deposition of both hydroxy apatite and the organic matrix. The result is the existence of rings of development, known as *annuli* which sometimes are used to estimate the fish age.

Bones, such as vertebrae and spines are not always easily distinguishable between species and they are lower in number than scales (e.g., 50 vertebrae/600 scales in sardine) [18, 19]; they consist of collagen hardened with inorganic salts [20]. Otoliths are only three pairs per fish, and are composed of aragonite and often affected by dissolution.

Fish scales found in sediments are essentially derived from scale shedding [21]. Aquarium experiments involving species such as sardine and anchovy show that most scales are lost during swimming and schooling, and that T type (T-typical) is the most abundant [21, 22]. The proportion of T scales to other scale types found in aquarium and sediments is always higher than its proportion in the respective body, which means that they are derived from differential shedding and not from individuals dying and settling to the bottom.

3. Where to find fish remains

The majority of works cited above is based in fish remains from small pelagic species, such as sardines, anchovies and herrings, mainly because they are widespread and abundant and are associated with centers of high biological productivity. Besides that, they are relatively short lived and have its recruitment dependent on environmental conditions, which make them a useful tool to evaluate the effect of environmental changes on fish recruitment.

Fish remains may be found in various places as archaeological sites (middens) and outcropped sediments [23], seabirds roosting and nesting sites, where guano is abundant, and principally, in marine sediments. However, the existence and abundance of fish remains on marine sediments is dependent on their degradation rates. It is consensual that low oxygen content and high sedimentation rate favour their preservation, which conditioned the sedimentary environments to search for. The ideal "scenario" for reconstruction of fish populations based in fish remains is to have "varves"or annual laminae (which help to obtain a good chronostratigraphy), abundant living small pelagic fish stock (which signifies abundant fish remains), low oxygen and high sedimentation rates. These exceptional conditions are not easy to find, which explains the very few places studied. Perfect examples can be given by the environment and sediments of Santa Barbara Basin (SBB) [7, 8, 9], Saanich and Effingham inlets [14, 15], and few other places (see works cited in the introduction). Off the Portuguese margin, sea bottom oxygen values are quite high (in the order of 6.5mg/l and 77% DOsat or larger) which make the Oxygen Minimum Zone (OMZ) incipient or even inexistent. The known sedimentation rates for the Portuguese shelf (determined only in the northern half) range from 0.07cm/year [24] to 0.57cm/year [25].

4. The study of fish remains

Sediment sampling for fish remains studies can be achieved with different approaches depending on the objective. If the objective is to study the last few decades or centuries, intact surface sediment sampling can be achieved with multi-corer and box-corer sampling devices, but if the objective is to reconstruct fish fluctuations of the last thousand years then a gravity corer, piston corer or vibrocorer must be considered. In the first POPEI survey it was sampled some gravity corers, ranging from 0.50 m to 1.86 m, as well as a vibrocorer of 5 m long from the Algarve's continental shelf (southern Portugal). Superficial samples were also collected by a Smith McIntyre grab from each station, for preliminary selection of cores study.

4.1 Methodology

The sediment study for fish remains begins with photo and Rx imaging analysis 'figure 1'. This is particularly important because it allows the definition and counting of the annual laminae or varves (alternation of layers of different densities related with seasonal sedimentation and thought to be equivalent to a single year) 'figure 1 (a)'. This work must be calibrated with sediment dating methods using isotopes like ^{14}C or ^{210}Pb. Gravity slumps and bioturbation can destroy or alter the continuity of laminated or banded sediments.

Once the internal core structure, number of laminae/varves, type of sediment, etc...is known, the criteria for subsampling can be established (per time-interval, per equal depth interval, per unit of mass, etc). The aim is to have the highest possible resolution per sample that yields enough material after sieving.

A first approach of this task was made before opening POPEI's cores, with Rx images. These, showed incipient or absent laminations, frequent bioturbation and relatively abundant shells 'figure 1 (b)', which make them very different from the sedimentary record of existing studies 'figure 1'.

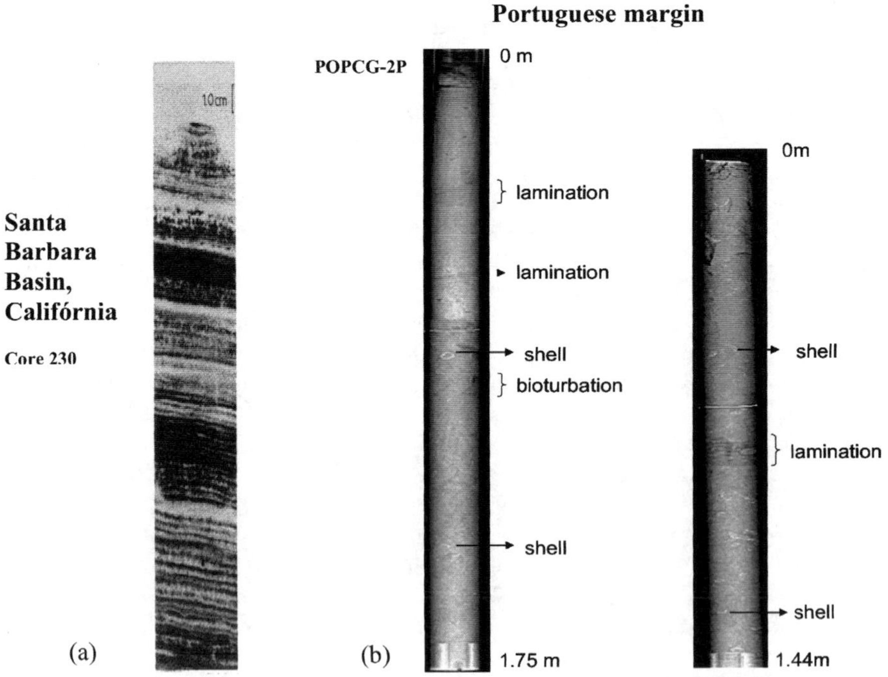

Figure 1. (a) Positive print of radiograph of Santa Barbara Basin Core 230 (dark and light laminations represent, respectively the winter (most dense) and summer (mainly composed by diatoms) sedimentation (from [8]); (b) Radiography of two closed cores, sampled off the Portuguese margin (POPCG-2P; POPCG-14P) with few perceptible laminations and some shells. In POPCG-2P bioturbation is also visible. (Core 230 is in a different scale from POPCG-2P and POPCG-14).

Laboratory sample procedure included a gentle sample treatment with diluted oxygen peroxide and sediment sieving through 355 μm and 125μm sieves. In the first sieve, the scales and the vertebrae of major dimension were retained while scale fragments and other vertebrae of minor dimension were found in the 125μm sieve. For both sieves the residual sediment was preserved in ethanol and observed and sorted with binocular microscope for scales and other fish remains. The identification of scales and bones is crucial for the correct ecological interpretation of their significance. Fish scales are easier to identify, because each species present different characteristics.

To help in this identification it is necessary to consult detailed identification criteria and to have good photos of fish remains collections. The existence of a regional atlas of fish remains is of utmost importance. Its elaboration must take into account the fact that scales morphology change according to different parts of the fish body 'figure 2'. Even with the help of such an atlas, the identification task is not easy; for example, to distinguish Pacific herring and sardine scales [14] and in general clupeioid scales [16] may result difficult.

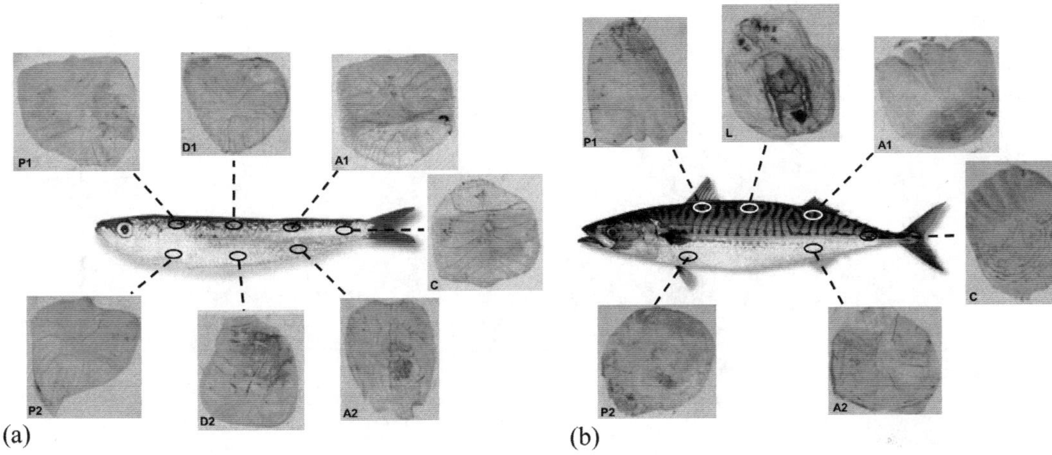

Figure 2. Fish body regions with the main types of scales (P1 – Pectoral dorsal, P2 – Pectoral ventral, D1 – Dorsal dorsal, D2 – Dorsal ventral, L – Lateral line, A1 – Anal dorsal, A2 – Anal ventral and C – Caudal) in sardine (*Sardina pilchardus*) (a) and mackerel (*Scomber scombrus*) (b) two of most abundant small pelagic fishes off Portuguese coast. These are photos from the POPEI fish scale atlas. The images of sardine and horsemackerel were taken, respectively, from http:// www.mackerel.cmima.csic.es/ictimed/cgi-bin/ictimed.cgi and http://www. Seafoodfromnorway. com/.

The presence/absence of fish remains in the sedimentary record does not mean necessarily the presence/absence of fish species. The species represented in the sedimentary record are those extremely abundant and from whom scales are readily lost. At the same time, transport processes could deposit scales far from their original shedding places and its abundance and/or proportion could be altered by destruction or dissolution.

The counting of fish scales follows criteria depending whether they are complete or not, and in the latter case, if it has the centre preserved or not. Depending on these observations, the scale can be counted as one or as half scale. The scales found in superficial sediments in the Portuguese continental shelf are very well preserved. In almost all of the observations they were found to be complete, thus allowing species identification 'figure 3'.

The relationship between scales and vertebrae (or bones) found in the different cores levels may be used to determine a remains preservation index, because it is supposed that the vertebrae suffer a longer degradation process than scales. Thus if the proportion scales/vertebrae decrease, it is a sign of scale destruction. The ratio between complete scales/fragments is also useful to estimate the scale degradation rate. In the POPEI project this ratio will be estimated as soon as data is available.

4.2. Data processing

Once fish scales are counted and identified, the next step is to calculate the Scale Deposition Rate (SDR). This parameter represents the flux of scales, expressed in number of scales per area per time interval, and could be used to estimate local fish abundances and relative abundances between species in a regional scale. However, this parameter has to be adapted if vertebrae or otoliths are used instead of scales. Computations for SDR estimations need dating methods.

SDRs of a 1700 year hindcast series of Pacific sardine and northern anchovy suggest that the fluctuations of these fish populations present a natural variability of higher magnitude than those suggested only by fisheries data [10] 'figure 3'. This variability usually expresses a pattern that could vary from decadal to centennial or even millennial scale [10, 26, 27].

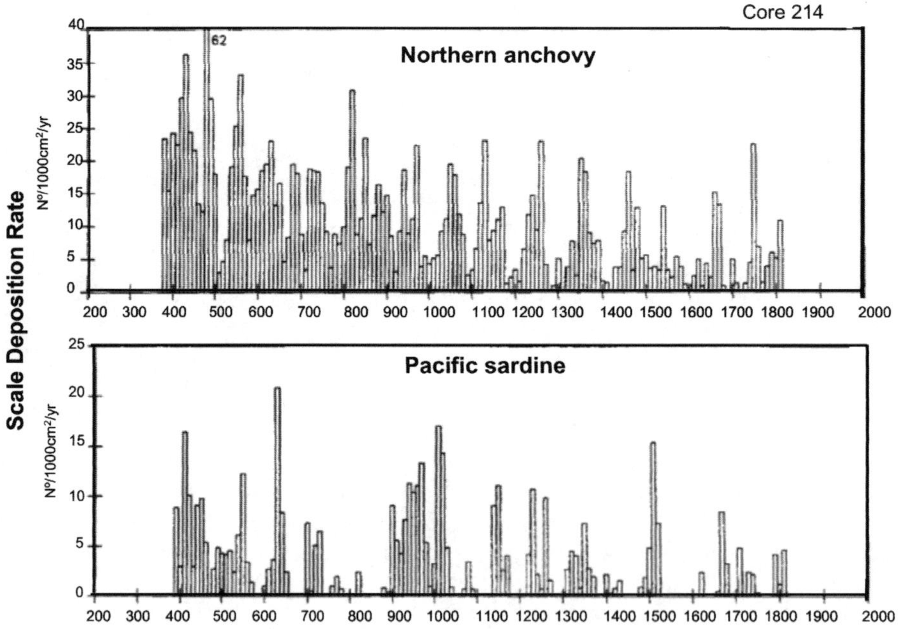

Figure 3. SDR's for Pacific sardine and northern anchovy reconstructed at 10 year-interval of a piston core (214) from Santa Barbara Basin. SDR's are given as mean annual values for each 10-year period (adapted from [10]).

The validity of the SDR parameter can be supported by several works. A good relationship was established between sardine SDR values and the biomass of fish younger than 2 years in the California Current, which is consistent with the observation that the majority of scales were from fish younger than 2 years [9]. This reinforces the idea that SDR is sensitive to the fish population living in the overlying water column [10]. SDR records determined by different investigators with independent chronologies showed that the decadal and centennial scale patterns of SDR were clearly coherent between cores [10]. SDRs of Pisco and Callao sedimentary records (Peruvian margin), distant from each other more than 300 km showed similar tendencies [28].

SDRs values are species specific and should be calibrated for each species and ocean region [29, 10]. This implies that SDR from sedimentary record have to be compared with other information like

IOP Conf. Series: Earth and Environmental Science 5 (2009) 012012 doi:10.1088/1755-1307/5/1/012012

fisheries historical data, even if it is known that this comparison will have errors from both sides. Fisheries data (landings) may not represent the "real" fish catches or may not reflect the local abundance of fish. The calculated SDR for a species might have an associated error from counting, identifying and measuring. However, it is expected that the tendencies are the same.

For this purpose, SDR of intact surface sediments acquires great importance and must always be determined. Calibrations of sardine and anchovy SDR to biomass have been done for Santa Barbara Basin [9, 10]. This was based on a 5 year sample resolution obtained from the analysis of box-cores and the data has demonstrated that SDRs values provide reasonable indices of changes in population biomass. It was found an equivalence of one scale per 400 000 tonnes of sardine and 200 000 tonnes of anchovy, respectively. In the Peruvian margin, at Callao and Pisco continental slope, SDR data and anchovy catch records indicate that they are fairly consistent [13]. In the varved sediments of Saanich Inlet, data from five box-cores allowed the conclusion that they reflect the major variations in the herring regional biomass [14].

The data obtained from fish remains must always be correlated with other productivity proxy's data such as diatoms, calcareous nannoplankton and foraminifera in order to verify how the fish populations variations are related with oceanic productivity changes. Cross-spectral analysis, for example, provide a non-dimensional measure between two time series and are particularly good for comparison off sedimentary scale records to other time series datasets [14].

5. Preliminary data of Portuguese margin

Preliminary data of POPEI project includes the analysis and identification of fish scales in superficial sediments. Some scales were also found in a rough observation of some core-catchers but they weren't identified. Some of the fish scales of superficial sediments belong to species as sardine 'figure 4 (a)' and mackerel 'figure 4 (b)' two of the most abundant fish species of the Portuguese coast. In general, they were well preserved. Also, a great number of vertebrae were observed in almost all the samples but until now none were possible to identify.

These are positive results attending to the fact that the Portuguese coastal environment is very different from the other sites studied, namely regarding the high level of oxygenation near the bottom.

A not so positive result is the fact that the cores present poor laminations, which will make detailed chronostatigraphy a harder task.

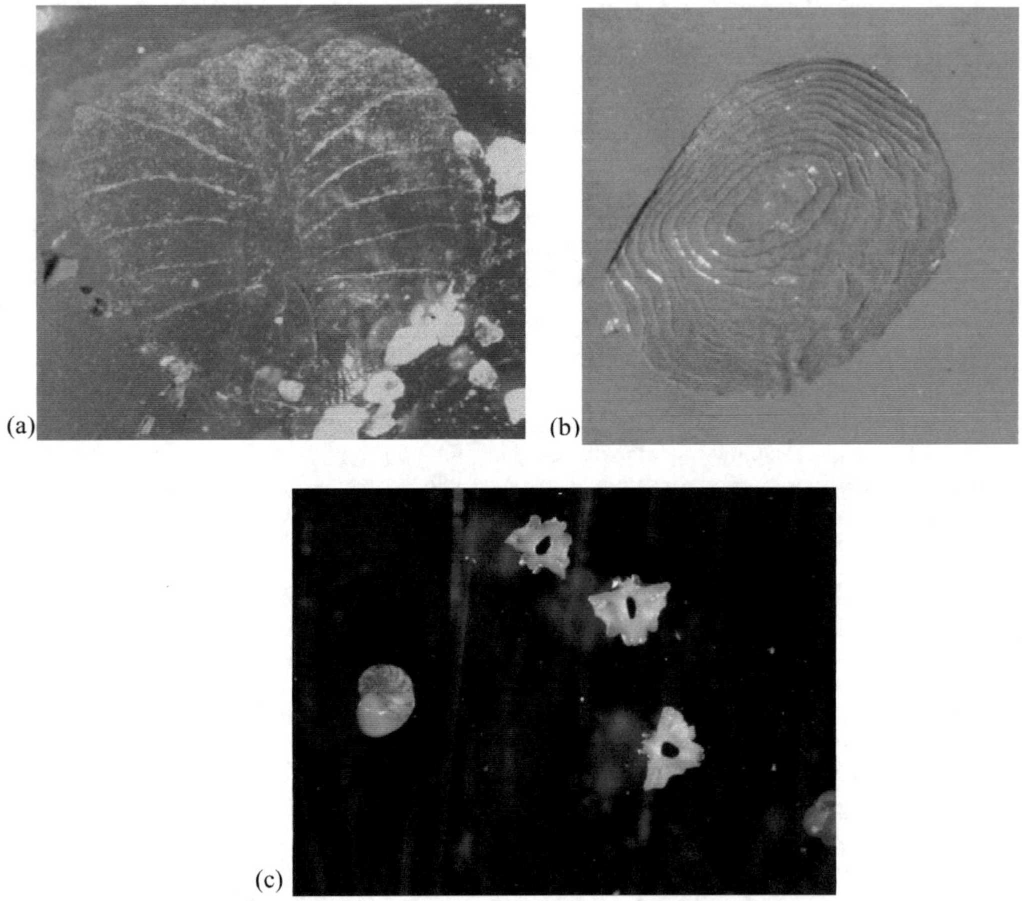

Figure 4. Examples of very well preserved fish scales and vertebrae found in Portuguese continental shelf sedimentary record. Fish scales of sardine (*Sardina pilchardus*) (a); mackerel (*Scomber scombrus*) (b) and vertebrae (c).

6. Conclusions

Studies of fish remains in the sedimentary record are the only way to determine the variability of fish populations beyond the range of fisheries records. In general, the obtained data presented in existing studies clearly suggest many different patterns of ocean and ecosystem variability that have not been observed directly in the 19-20th century.

Fish remains studies could provide information regarding long-term variability in pelagic species composition and their relative abundances. When assembled with other productivity proxies, they can be connected with oceanic productivity changes and may contribute to the determination of the forcing factors for ocean productivity changes, namely oceanographic conditions.

Preliminary results of the POPEI project for the southern Portuguese coast show that well preserved scales can be found in well oxygenated environments.

Acknowledgments

This paper was undertaken as a part of the research project POPEI (PDCT/MAR/55618/2004) funded by Fundação para a Ciência e Tecnologia (Portugal). The authors wish to thank GRICES that provides financial support under the Portuguese-Mexican cultural agreement that allow the visit of Vicente Ferreira-Bartrina to Portugal in the context of the cited project. Thanks are also due to Renato Salvatteci and Francisco Fatela, the two reviewers, for the careful review and constructive comments of the initial manuscript and to all the partners of the project from different institutions (Instituto Hidrográfico, Faculdade de Ciência da Universidade de Lisboa e Instituto do Mar) for valuable discussions.

References

[1] Kawasaki T 1991 *Long-term variability of pelagic fish populations and their environment* (Oxford: Pergamon Press) pp 47–60

[2] Lluch-Belda D, Crawford R J M, Kawasaki T, MacCall A D, Parrish R H, Schwartzlose R A and Smith P E 1989 World wide fluctuations of sardine and anchovy stocks: the regime problem *S. Afr. J. Mar. Sci.* **8** 195–205

[3] Alheit J and Hagen E 1997 Long-term climate forcing of European herring and sardine populations *Fish. Oceanogr.* **6 (2)** 130–9

[4] Borges F, Santos A M P, Crato N, Mendes H and Mota B 2003 Sardine regime shifts off Portugal: a time series analysis of catches and wind conditions *Scientia Mar.* **67** 235–44

[5] Santos A M P, Kazmin A S and Peliz A 2005 Decadal changes in the Canary upwelling system as revealed by satellite observations: Their impact on productivity *J. Mar. Res.* **63** 359–79

[6] Santos A M P, Borges F and Groom S 2001 Sardine and horse mackerel recruitment and upwelling off Portugal *ICES J. Mar. Sc.* **58** 589–96

[7] Soutar A 1967 The accumulation of fish debris in certain California coastal sediments, *CalCOFI Reports* **11** 136–9

[8] Soutar A and Isaacs J D 1969 History of fish populations inferred from fish scales in anaerobic sediments off California *CalCOFI Reports* **13** 63–70

[9] Soutar A and Isaacs J D 1974 Abundance of pelagic fish during the 19th and 20th centuries as recorded in anaerobic sediment of the Californias *Fish. Bull.* **72** 257–73

[10] Baumgartner T R, Soutar A and Ferreira-Bartrina V 1992 Reconstruction of the history of Pacific sardine and northern anchovy populations over the last two millennia from sediments of the Santa Barbara Basin, California *CalCOFI Rep.* **33** 24–40

[11] Holmgren-Urba D and Baumgartner T R 1993 A 250-Year history of pelagic fish abundance from anaerobic sediments of the central Gulf of California *CalCOFI Rep.* **34** 60–8

[12] DeVries T J and Pearcy W G 1982 Fish debris in sediments of the upwelling zone off central Peru: a late Quaternary record *Deep-Sea Res.* **28** 87–109

[13] Salvatteci R, Gutierrez D, Field D, Ferreira V, Velazco F, Sifeddine A, Baumgartner T, Ortlieb L and Soler P 2006 Fluxes of scales and other fish remains en marine sediments in front of Callao, Perú en the last 550 years *XIII Cong. Geol., Soc.Geol.of Peru*, Lima, Perú, Oct.

[14] O'Connell J M and Tunnicliffe V 2001 The use of sedimentary fish remains for interpretation of long-term fish population fluctuations *Mar.Geol.* **174** 177–95

[15] Patterson R T, Prokoph A, Wright C, Chang A S, Thomson R E and Ware D M 2004 Holocene solar variability and pelagic fish productivity in the NE Pacific *Palaeont.Electr.* **7** 17

[16] Shackleton LY 1986 *An assessment of the reliability of fossil pilchard and anchovy sales as fish population indicators off Namibia* MSc thesis University of Cape Town p 141

[17] Wallin O 1957 On the growth structure and developmental physiology of the scales of fishes *Rep.Inst.Freswater Res.* (Drottningholm: Inst.of Freswater Res) **38** 385–477

[18] Andreu B 1969 Las branquispinas en la caracterización de las poblaciones de Sardina pilchardus (Walb) *Invest. Pesq.* **33** 425-607

[19] Bateson W 1894 *Materials for the study of variation treated with especial regard to discontinuity in the origin of species* MacMillan and Co., London, p 598
[20] Wheeler A and Jones AKG 1989 *Fishes* (Cambridge: Cambridge University Press) p 210
[21] Shackleton L Y 1988 Scale shedding: an important factor in fossil fish studies *J. Cons. Int. Explor. Mer.* **44** 259–63
[22] Lozano H 1997 *Análisis del uso de escamas fósiles de la cuenca de Santa Bárbara, California, para reconstruir variaciones en abundancia y estructura de las poblaciones de sardina y anchoveta* MSc thesis. Centro Invest.Cient.Educ.Sup.Ensenada (CICESE) Ensenada Mexico p 97
[23] Fitch J E 1969 Fossil records of certain schooling fishes of the California Current System, *CalCOFI Rep.* **13** 71–80
[24] Jouanneau J M, Weber O, Drago T, Rodrigues A, Oliveira A, Dias J M A, Garcia C, Schmidt M and Reyss J L 2002 Present day sedimentation and sedimentary budgets on the northern iberian shelf *Progr. in Oceanog.* **52** 2-4 261-75
[25] Drago T, Oliveira A, Magalhães F, Cascalho J, Jouanneau J M and Vitorino J 1998 Some evidences of northward fine sediment transport in the northern Portuguese continental shelf Oceanol. Acta **21** 2 223-31
[26] Baumgartner T R, Struck U and Alheit J 2004 GLOBEC investigation of interdecadal to multi-centennial variability in marine fish populations *PAGES News* **12(1)** 19–21
[27] Tunicliffe V, O'Connell J M and McQuoid M R 2001 A Holocene record of marine fish remains from the Northeastern Pacific *Mar. Geol.* **174** 197–210
[28] Gutiérrez D, Field D, Salvatteci R, Baumgartner T, Soutar A, Díaz Ochoa J, Velazco F, OrtliebL, Sifeddine A, Tapia P, Lange C B, Ferreira V and Vargas G 2006 Decadal to centennial variability of the Peruvian upwelling ecosystem during the last centuries as inferred from fish scale deposition rates of anchovy and other marine sediment records *Int. Conf.Humboldt Current System*, November
[29] Shackleton L Y 1987 A comparative study of fossil fish scale from three upwelling regions *Sout.Afr. Jour.Mar.Sci.* **5** 79–84

Institute of Physics
Dirac House, Temple Back
Bristol BS1 6BE UK

CURRAN ASSOCIATES INC
proceedings
.com